Energy, Climate and the Environment

This is a critical time for nature-society relations as the global community faces the challenge of making food, energy, and transport systems more sustainable and resilient. Energy, climate and environmental systems are central to staying within planetary boundaries amid rapidly changing social, economic, political, and environmental conditions. This book series will focus on these topics, their interlocking impacts, interactions and uncertain outcomes. Contributions are encouraged from emerging and established scholars working within and/or across the energy, climate and environmental domains. The series provides an outlet for scholarship, debate, discussion, and dialogue between social sciences, humanities, physical and applied sciences. Its objective is to promote interdisciplinary and transdisciplinary research, whilst simultaneously fostering positive social change. The series provides an opportunity for scholars to connect state of the art science with the policymaking and practitioner community, accelerating the transition to a resilient future.

Tracey Skillington · Annalisa Setti
Editors

Climate Change Resilience Across Societal Contexts

Challenges, Potentials and Visions for the Future

palgrave
macmillan

Editors
Tracey Skillington
University College Cork
Cork, Ireland

Annalisa Setti
Psychology, Enterprise Centre
University College Cork
School of Applied Psychology
Cork, Ireland

ISSN 2947-8561 ISSN 2947-857X (electronic)
Energy, Climate and the Environment
ISBN 978-3-031-95019-3 ISBN 978-3-031-95020-9 (eBook)
https://doi.org/10.1007/978-3-031-95020-9

Cover credit: Scott Dumas/Alamy Stock Photo

This Palgrave Macmillan imprint is published by the registered company Springer Nature Switzerland AG
The registered company address is: Gewerbestrasse 11, 6330 Cham, Switzerland

In loving memory of Liana and Giacomo Setti

Acknowledgement

This research has been supported by the European Union under the Horizon Europe research framework under grant agreement no. 101137209. The sole responsibility for the content of this website lies with the GoGreen Next project and in no way reflects the views of the European Union.

Contents

Notes on Contributors

Jasmine Byrne, BA, MA is an independent researcher.

Denise Cahill is coordinator of Cork WHO Healthy Cities, an interagency initiative aimed at addressing the social and environmental determinants of public health.

Marica Cassarino is an Environmental Psychologist at University College Cork. Using the Sustainable Development Goals as framework, her research interests are in socio-ecological perspectives on environmental attitudes and behaviours, as well as environmental impacts on wellbeing, particularly for vulnerable groups. She is passionate about exploring these topics through community engaged work and cross-sectoral and interdisciplinary collaborations.

Roberto Cibin is a social scientist at University College Cork. His research delves into the interplay between social and technical dimensions in the design of technologies, platforms and services. He places particular attention on addressing inequalities, environmental issues, and power dynamics within these contexts. He employs participatory and community-based approaches to foster empowerment, inclusion, responsibility, and sustainability in design processes.

Mirco Fasolo is a Full Professor of Developmental Psychology at the G. d'Annunzio University of Chieti-Pescara, specializing in language development in typical and atypical populations. His research explores linguistic input, prosody, and neurophysiological methods such as biofeedback and EEG. He also coordinates the Level II Master's program on Neurodevelopmental Disorders and serves on several scientific boards focused on specific learning and communication disorders.

Dr. Pádraig Flattery is Senior Climatologist at Met Éireann with expertise in climate change research, strategic development, and data visualization.

Elaine Gallagher is based at the Innovations Value Institute, Maynooth University, Maynooth, Ireland.

Dr. Amin Sharifi Isaloo is a lecturer in the Department of Sociology and Criminology at University College Cork. His fields of interests include race, ethnicity, migration, nationalism, development and globalisation, politics, religion, arts and culture focusing on sociological and anthropological interpretations of symbols, images and ritual performances. He is the author of the book 'Power, Legitimacy and the Public Sphere: The Iranian Ta'ziyeh Theatre Ritual' (2017). His recent peer-reviewed publications are 'Liminality in the Direct Provisional system '(2020), 'Liminality and Modern Racism' (2021), 'Racism and the far right' (2023), and the co-editor of upcoming book entitled 'Modern racism and the far right'.

Johanna Marie Kirsch is a sociologist whose research interests include right-wing populism, conspiracy beliefs, and social and gender inequality. She has published on how global care chains contribute to gendered inequality and economic marginalisation. She also collaborated with Tracey Skillington on an Irish Research Council-funded project exploring human-nature relations in urban environments and the justice dimensions of unequal access to urban nature.

Sara Lind holds a Master's degree in Human Rights and Democracy (2023). She has a strong interest in socially sustainable development, with a particular focus on Indigenous rights and child rights. She currently works in the civil society sector, focusing on children's rights in underprivileged socio-economic areas.

Francesca Lionetti, Ph.D., is a psychologist and psychotherapist, and an Associate Professor of Developmental Psychology at the University of Pavia, Italy. Her research explores Environmental Sensitivity and Sensory Processing Sensitivity across the life span, with a particular focus on parenting and children's socio-emotional development. She is also a co-founder of sensitivityresearch.com, a platform dedicated to advancing the scientific understanding of sensitivity.

Tadhg E. MacIntyre Ph.D. is an Associate Professor of Environmental Psychology at Maynooth University, Department of Psychology and Innovation Value Institute. He is a funded investigator with the Research Ireland Insight Centre for Data Analytics. He currently leads two EU Horizon Projects GoGreen Routes and GoGreen Next (2024–2028) both focused on urban health and well-being. He is also a work package lead on Green Talent (Horizon Europe project) which is a widening action to build capacity in supporting nature-based solutions across Europe.

Niamh McCarthy is an Assistant Psychologist working across two clinical teams: Adults with Intellectual Disabilities (ID) and the Children's Disability Network in Kare. She co-facilitates child and adult psycho-educational groups and individual intervention around anxiety, Autism and emotional well-being. As part of the Quality of Life sub-team for adults with ID she is conducting the quantitative analysis and dissemination of findings of the quality of life survey. Additionally, she works as a research assistant on the *chAnGE* project with University College Cork where she is contributing to designing a micro-credential on mental resilience to climate change in ageing for health and social care professionals.

Darren McCauley is Professor in Environmental & Social Justice, Newcastle Law School, Newcastle University, UK. Prof. McCauley was previously Chair in the Management of International Social Challenges at Erasmus University Rotterdam in the Netherlands. He maintains a Visiting Professorship there as a part of the Science, Public Issues and

Imaginaries group of DPAS. Darren held his Senior Lectureship at the University of St. Andrews where he led the university's sustainability institute and Arctic Research Centre. He notably held lectureships at Stirling, Queens and Trinity College Dublin. Prof. McCauley also holds further visiting professorships at several institutions and many external commissions of trust (editorial and editorships, peer review colleges, external boards).

Patrick McGurn has worked with farmers on the Aran Islands for 15 years, designing and implementing programmes to enhance farmland biodiversity. He was the project manager for AranLIFE and the follow up project, Caomhnú Árann. Patrick has a background in agriculture and ecology. After graduating in Agriculture from Queen's University Belfast, he went on to work in the agriculture advisory services during which time he completed a Ph.D. on species-rich grasslands in Fermanagh. He is presently with ACRES BurrenAran Co-operation Project (BUA CP), working with farmers in managing and conserving the ecological, archaeological and cultural heritage of the Islands.

Dr. John McNamara is Senior Health and Safety Specialist Advisor with Teagasc. He previously was both an agricultural College Trainer and Principal. He holds a Master's Degree in Agricultural Science and a Diploma in Occupational Safety, Health and Welfare at Work along with a Ph.D. which examined health and safety adoption by farmers based on a risk assessment document approach. He is an Adjunct Associate Professor at the School of Agriculture and Food Science, UCD.

Niamh Ní Chongahile is a registered occupational therapist with training in yoga and nature-based interventions. Niamh is currently completing her M.Sc. in Environmental Psychology at Maynooth University. Previously, Niamh was the national lead for the Woodlands for Health program working for the NGO Mental Health Ireland.

Dr. Tom O'Dwyer is Head of the Teagasc Climate Change Signpost programme assisting farmers to adopt climate change mitigation measures. He previously worked as Head of Knowledge Transfer for Dairying, a county Advisory Services manager and as an Advisor/ Project Specialist. He holds a Bachelor of Agricultural Science and MBA degrees along with a Doctorate of Business Administration (DBA) which focused on the areas of self-leadership and psychological empowerment.

Dr. Johanna Ohlsson (Ph.D. Ethics), Senior Lecturer in Human Rights and Democracy at University College Stockholm and Postdoctoral Researcher in LU4Water at the Division for Human Rights Studies at the Department of History, Lund University. Her research focuses on the role, understanding, and function of human and Indigenous rights, theories of justice and justification, and issues of just sustainability and resilience in the Arctic.

James O'Rourke , BA, HDip is graduate student and assistant psychologist. His research interests include imagination, psychosocial and socio-cultural aspects of climate change inaction, and neurodevelopmental differences in children.

Dr. Giovanni Ottoboni is a chartered psychologist with a Master in Clinical and Experimental psychology and a Ph.D. in Psychological Science. His research interests include those aspects of human cognition that supervise action imitation, human-object interaction and body representations. He is involved in research about ageing, work retirement, social inclusion and loneliness, and ageing and technology. He co-designed an online curriculum of study in psychological care in dementia. More recently he expanded his research to the professional ethics of psychological practices. He teaches Personality and Cognitive Styles at the Department of Psychology, University of Bologna, and Nutrition and Psychological and Cognitive Well-being at the Department of Quality of Life, University of Bologna, Italy.

Stephen Przybylinski is Assistant Professor of Geography, Environment, and Spatial Sciences at Michigan State University. He is an urban and political geographer with interests in homelessness, property, justice, law, and citizenship. He is the author of *The Injustice of Property: Homeless Encampments and the Limits of Liberalism* (University of Georgia Press) and co-editor of *Theorising Justice: A Primer for Social Scientists* (Bristol University Press).

Kevin Ryan is the Executive scientist of the Air Quality Unit within Cork City Council. He is leading the development and implementation of the Cork air quality strategy.

Sakshi is a Lecturer in Law and Social Justice at the Newcastle Law School. Her research areas include comparative environmental law, constitutional law, Marxist legal theory and political economy, sovereignties, and jurisprudence. Sakshi's research is transdisciplinary with a deep situatedness in law. She is currently working on her new project, Plural Sovereignties and the Emergence of New Legalities.

Dr. Annalisa Setti, Ph.D., CPsychol, is Senior Lecturer/Associate Professor at the School of Applied Psychology and affiliated Principal Investigator at the Environmental Research Institute, University College Cork, Ireland. She has an established track record on the psychology of ageing. Her current research interest is at the intersection between individual (level of sensory sensitivity, nature connectedness) and environmental factors (exposure to positive or adverse nature experiences) in determining human cognition, wellbeing, and pro-environmental action. Her research is funded by Horizon Europe, Marie Sklodowska Curie (EU), Cities Mission (EU), Research Ireland and, previously, the Health Research Board and the Irish Research Council. She collaborates with Cork City Council and Cork-Healthy Cities to support achievement of the goal of 2030 climate neutrality in Cork.

Tracey Skillington is a Senior Lecturer/Associate Professor and Head of Department of Sociology & Criminology, University College Cork. Her research focuses on a broad range of issues relating to climate justice and democratic theory. She has published in a variety of international journals, including the European Journal of Social Theory, International Journal of Human Rights, the British Journal of Sociology and has two monographs published to date, *Climate Justice & Human Rights* (2017, Palgrave) and *Climate Change and Intergenerational Justice* (2019, Routledge).

Alessandra Sperati is a developmental psychologist and postdoctoral researcher at the University G. d'Annunzio of Chieti-Pescara, Italy. Her research explores individual differences in Environmental Sensitivity and its impact on development. Her work includes developing assessment tools for children and exploring neurophysiological correlates in both children and adults. She is also interested in parenting and attachment, and how sensitivity interacts with environmental influences in socio-emotional adjustment across life.

Maria Spinelli, Ph.D., is a psychologist and psychotherapist, and an Associate Professor of Developmental Psychology at the University G. d'Annunzio Chieti-Pescara, Italy. Her research focuses on parenting, child development, and individual differences, with a particular emphasis on cultural adaptation and well-being.

Andrea Stitzel has a Masters Degree in Psychology and is currently pursuing her Ph.D. in medical sciences, focusing on climate resilience of primary care professionals. Her research interest lies in climate literacy of health care professionals. She has been involved in several EU research projects on climate change, health and ageing. Recently, she designed an 11 ECTS Bachelor module on Planetary Health Management in which she is the main lecturer as well as responsible for the overall module coordination. In addition, she is involved in the development of micro-credentials for Bachelor and Master level health care students in the area of climate change and health. She teaches at the Department of Health Sciences and Social Work at Carinthia University of Applied Sciences primarily in the study programme Healthcare Management.

Raegan Thompson is a Psy.D candidate in the field of Clinical Psychology at the National Louis University. A graduate of psychology and religion from the New College of Florida in 2024, Raegan was a Summer Intern on the GoGreen Next project at Maynooth University. Her research interests are in urban health, nature-based interventions and mental health including supporting those experiencing personality disorders.

Reetta Toivanen Legal anthropologist by training, (Ph.D. Humboldt University in Berlin) is since 2018 full professor in Sustainability Science (Indigenous Sustainabilities) at the Helsinki Institute for Sustainability Science (HELSUS) and Department of Cultures at the Faculty of Arts at the University of Helsinki, Finland. She is also a docent in social and cultural anthropology. She is the vice-director of the Centre of Excellence in Law, Identity and the European Narratives (EuroStrorie) funded by the Academy of Finland (2018–2025) leading the Subproject Migration and the narratives of Europe as an "Area of freedom, security and justice". She is a non-resident senior research fellow at the European Centre for Minority Issues (ECMI) in Flensburg (Germany). 2023–2025 she is a member of the Finnish Government's Expert Panel on Sustainability.

Her recent publications include: "Legal Anthropology". In M. Lindroth, H. Sinevaara-Niskanen, & M. Tennberg (Eds.), Critical Studies of the Arctic: Unravelling the North. Palgrave Macmillan. (2023); "Protecting Indigenous identities? An example of cultural expertise on Sámi identity". Legal pluralism and Critical Social Analysis, 54(2–3) (2023); "Narratives on 'refugee knowledge' in the institutions of Europe" (with B. Bergholm). In M. Kmak, & H. Björklund (Eds.), Refugees and Knowledge Production: Europe's Past and Present, Routledge (2022).

Dr. Angela Veale is Senior Lecturer in the School of Applied Psychology, UCC. Her research and publications focus on the psychosocial effects of violence, post-conflict social reintegration, migration and globalisation on children, youth and families.

Dean S. Venables is a Senior Lecturer in chemistry at University College Cork. His research interests include air quality, low-cost sensors, and developing spectroscopic instruments to detect trace gases and aerosols. His current projects focus on nitrogen dioxide from vehicles, agricultural releases of ammonia, and the use of low-cost sensors to understand the spatial and temporal characteristics of air pollution in Ireland.

Vladislava Vladimirova is an Associate Professor in Cultural Anthropology and a Senior Lecturer in Eurasian Studies at the Institute for Russian and Eurasian Studies (IRES) and the Department of Cultural Anthropology and Ethnology at Uppsala University. She is a cultural anthropologist doing long-term research in Circumpolar Eurasia. Her main interests are in the field of Indigenous rights and ethnicity, environmental governance and justice, gender, moral economy, and conservation of environmental and cultural heritage in the Arctic.

Corine Wood-Donnelly is a Professor of Social Science with expertise in International Relations and the High North at Nord University. She has regional expertise in the Arctic with substantive knowledge of governance frameworks, maritime law, and state practices. Her research interests include normative evolutions in sovereignty and governance, maritime search and rescue, territorial and resource sovereignty, performative geopolitics, and the role of values in decision-making. As a scholar of the Arctic, Corine conducts interdisciplinary work within International Relations and political geography, also drawing from public international law, history and theories of justice.

List of Figures

1

Introduction

Tracey Skillington and Annalisa Setti

In this edited volume, a multidisciplinary team of researchers recount the findings of original field research exploring lived experiences of climate change amongst cohorts residing in various regions of the world. A diverse range of lived experiences of climate change impacts are reported upon and analysed to provide a clearer picture of how the goal of climate change resilience is regularly complicated by a series of challenges. That is, psychological, inter-personal, socio-cultural, political, and economic challenges that disrupt prospects for an equitable and peaceful transition to a green sustainable future. Aligning with Bronfenbrenner Ecological Systems Theory (1979), we conceptualise the complex network of relationships individuals and wider collectives

T. Skillington (✉)
School of Society, Politics and Ethics, University College Cork, Cork, Ireland
e-mail: t.skillington@ucc.ie

A. Setti
School of Applied Psychology, University College Cork, Cork, Ireland
e-mail: a.setti@ucc.ie

T. Skillington and A. Setti (eds.), *Climate Change Resilience Across Societal Contexts*, Energy, Climate and the Environment,
https://doi.org/10.1007/978-3-031-95020-9_1

operate in, noting how resilience develops as a layered phenomenon, where psychological and social responses to climate change adversity and actions are co-determined by the opportunities afforded by the surrounding environment and culture (Skillington, 2023; The Lancet, 2020). Of relevance here are immediate relationships with the family, community support structures, neighbourhoods, cities and their governance systems, as well as wider socio-cultural, political, and economic contexts. Within these various settings, climate change impacts add to existing burdens and exacerbate many societal inequalities.

The chapters of this volume set out to explore these relations and their impact on prospects for effective resilience building for the future. With its emphasis on detailed empirical research, this research responds to the call of various bodies, including the American Psychological Association (2023), the World Health Organization (2025), and the IPCC (2022) for a more intimate engagement with the impacts of climate change on everyday lives, as well as a more open dialogue on whether strategies for realising various Sustainable Development Goals (e.g. Goal 10 on Reduced Inequalities and Goal 13 on Climate Action) usefully address the concerns of those most affected. With a heavy emphasis in official policy discourse on the need for communities to abandon many established socio-cultural practices and adjust comprehensively to a series of new climate imperatives (i.e. changing social, environmental, economic, technological, and cultural commitments), the danger is that the potential risks these demands pose to the health and wellbeing of vulnerable cohorts will be overlooked. Greater consideration, therefore, should be given to the way different groups experience climate change related disturbance, envisage their present circumstances and future prospects, define preferred approaches to climate change adaptation and mitigation, and understand the risks posed by climate change.

This volume assesses how these imperatives affect communities living at the cold face of a climate changing world, noting how they can further undermine the health and wellbeing of many (ACOSS, 2015) and diminish prospects for an inclusive programme of resilience-building and effective climate change mitigation. At its core, an equitable approach to resilience-building offers a comprehensive appraisal of the ways current climate actions and sustainable development policies are failing communities. There is considerable evidence to suggest that policy still lags behind research in terms of efforts to address factors currently threatening a viable 'wellbeing economy' (European Council, 2019).[1] Inclusivity remains a key element missing in effective SDGs realisation. Contributors to this edited volume consider these issues in depth, drawing on the findings of funded research to assess factors that currently heighten climate change marginalisation, the risk of displacement, increase levels of stress, anxiety, and grief amongst those who encounter climate change as a loss of traditional ways of living, livelihood, good health, community connection, security of place, cultural heritage, and more.

Highlighting significant gaps in the current policy applications, this volume also brings examples of resilience, whereby different communities or groups have demonstrated the ability to adapt to uncontrollable circumstances, while remaining cohesive and future oriented. The strengths organically displayed by these groups could be used more effectively to inform wider policy and practice.

Contributors bring expertise from a variety of disciplinary fields, including geography, philosophy, sociology, psychology, anthropology, law, public health, and international development, to examine factors weakening the adaptive capacities of populations in various world regions, before considering how these elements might be addressed and a stronger people centred approach to resilience established.

Contributors explore contexts where increasing exposure to climate change adversity provides important empirical settings of learning, where more situationally grounded resilience and recovery techniques are deployed by communities attempting to adjust to changing planetary conditions (e.g. higher risks of flooding, prolonged drought, intense heatwaves, degraded air quality, wildfire risks, storm surges, etc.) and risks of displacement and largescale destruction. Climate change events have proven to be an important stimulus to action, encouraging 'transformative work' (Beck, 2015) in communities' preparedness for disaster (through skills learning and new cooperative initiatives). Frequently, working to create safer 'common futures' is inspired by memories of overcoming adversity and enabling flourishing to occur against all the odds.

Our aim, therefore, is to also assess what motivational powers community heritage offers in the struggle to preserve a good life in increasingly challenging ecological and social circumstances. Likewise, the chronic effects of climate change on different groups should not be underestimated. Even when the threat of destruction seems to be relegated to a relatively far future, as is the tendency in many regions of the Global North, the chronic threat of climate change requires far reaching efforts at resilience, adaptation, behavioural and attitudinal change. Therefore, this volume aims to speak to communities and policy makers operating in lower as well as higher risk areas in the hope of stimulating further cross-regional and cross- institutional learning amongst different groups, cultures and geographical locations.

The volume is divided into four sections, with each section focusing on one category of 'population of concern' (e.g. US Global Climate Change Research Program (2023) 'Populations of Concern'). The empirical research documented here is drawn from areas as diverse as the Middle East, Central Europe, the United States, the UK, Ireland, and various regions of the Arctic. Chapters will explore levels of exposure to climate risks across different cohorts and world regions and consider what role variables such as age, ethnicity, socio-economic background, gender, culture, occupation, and political participation play in inhibiting or, indeed, boosting social, behavioural, psycho-emotional and civic capacities to adapt effectively to newer climate challenges.

Chapters the First section of this volume assess how climate change impacts are felt by different urban populations, noting whether diminishing access to wild nature and designated green and blue public spaces disrupts important protective behaviours (e.g. stress reduction, sensory stimulation, the attention restoration benefits derived from regular immersion in green and blue 'therapeutic landscapes') and negatively impacts the social, psychological, cognitive, and physical well-being of residents (e.g. diminished air quality, higher levels of noise and light pollution, more frequent exposure to heat, drought, flooding, etc.).[2] The analysis explores how urban residents negotiate these changes as they affect their everyday lives, asking if links between decreasing levels of direct contact with a clean, flourishing nature, and a decline in residents' health and wellbeing can be minimised. This section also aims to bring to the fore other individual or social affecting the health and wellbeing of some more readily than others. Contributors consider the efforts of city resident groups to reclaim aspects of their cities' ecological and cultural heritage and restore health + life + work + play balance in ways that centrally address deficiencies in the distributive, recognition and procedural justice dimensions of existing sustainable development programmes for their city.

In his exploration of the experiences of residents of sanctioned homelessness encampments in the city of Portland, Oregon in the US, for instance, **Przybylinski in Chapter2** documents how climate change effects and pollution negatively impact the lives of the city's homeless populations. Portland has encountered unprecedented temperature highs in recent years resulting in a significant loss of life (older residents, the homeless and those on low incomes who accounted for most reported heat-related deaths in Oregon in 2023, see Climate and Health in Oregon 2023 Report). The failure of government to provide adequate care to the city's homeless during heatwaves or storms has resulted in homeless encampments becoming spaces of high-risk economic, social, and environmental inequalities. That is, spaces where some of the cities most at risk populations are exposed to harm that are largely avoidable if the necessary precautionary measures were provided. The growing regularity of extreme weather events, such as wildfires, heatwaves and storms, means that the numbers displaced continue to rise as do the numbers seeking refuge in homeless encampments. Przybylinski notes a tendency for encampments for the homeless to be located near motorways, airports, industrial waste or dumping sites. The latter become 'sacrifice zones' (Lerner, 2010) where camp residents are knowingly subjected to high levels of environmental pollution with detrimental health impacts. Przybylinski considers the unacceptability of this arrangement from a justice perspective and its incongruence with effective resilience-building measures proposed for climate smart, 'inclusive cities' (see UNEP, 2024).

In **Chapter3, Skillington and Kirsch** call for a more critical sociological appraisal of the '15-minute' climate smart city' model. In particular, the failure of government to take the needs and experiences of more marginal communities into consideration when planning and implementing sustainable development plans for cities such as Cork in the south of Ireland (designated a climate smart city by the European Commission in 2022). Research conducted by the author's points to disparities in the availability of accessible green and blue spaces in the city. As a consequence, the restorative benefits of its nature are unevenly distributed. Beyond lifestyle (e.g. sedentary lifestyles, poor diet, smart phone addiction, etc.), other social factors, the authors argue, also play a role in shaping the rate of 'extinction of experiences of nature' (Soga & Gaston, 2016). Embedded inequalities in quality of access (due to poor transport, infrastructural design, heavy traffic, and safety concerns) to the city's green and blue spaces have been shown to limit vulnerable cohorts' abilities to avail of nature's recuperative advantages. Similarly, in **Chapter4, Cassarino, Cibin, Venables, Cahill & Ryan** note constraints in social and structural urban contexts that disrupt the capacities of disadvantaged communities to move away from polluting environments. The authors focus on the issue of air pollution and the positive efforts of city authorities to date to monitor its impacts on health and wellbeing and foster behavioural change.

However, Chapters 2, 3, 4, and 5 all identify challenges which continue to limit the equity factor in current 'climate smart' city living and consider how greater efforts could be made to strengthen the deliberative capacities of present decision-making and address persisting structural inequalities in 'inclusive' green urban designs.

In the second section of this volume, contributing chapters focus on the issue of resilience-building across generations. While everyone is affected by climate change (e.g. flooding, heat waves, biodiversity loss, reduced mobility due to climate events, pandemics), people at different stages of life experience such changes with specific psychological impacts. The chapters in this section propose frameworks to address these issues and account for initiatives that aim to foster resilience building amongst elders and youth on their own terms, and in ways that enhance quality of life and independence.

In **Chapter6, Setti, Stitzel, Ottoboni & McCarthy** explore how older adults can build resilience by looking at research on ageing well and the impact of the lived environment. By examining different frameworks, they stress that ageing is not a fixed process or experienced as a constant decline. As a consequence, resilience-building efforts should focus on the individual's current situation, their perspective, and future goals to avoid paternalism or ageism. As the authors highlight, older adults are not a homogeneous group but display notable differences in neurological responses, cognitive and physical health, psychological and socio-cultural dimensions which all play an important role in resilience-building. The authors review existing research on resilience in ageing and suggest ways to improve both the theories and how they are applied in practice. Micro and macro resilient promoting actions, including coping strategies, and practical changes to home or lifestyle, stem from older people focusing on self-empowerment. Promoting nature connectedness is a way to foster resilience and promote pro-environmental behaviours, especially (but not only) for those who are neurologically particularly sensitive to environmental influences. Climate action is also considered as an avenue to combat solastalgia, i.e. the distress or sadness people feel when their familiar environment is damaged or changed, and climate change anxiety, i.e. feelings of worry, fear, or stress about the impacts of climate change on the environment, society, and an uncertain future.

In **Chapter7, Veale, O'Rourke & Byrne** assess symptoms of climate maladaptation and eco-anxiety amongst contemporary youth (aged 18–30), noting the ways in which these conditions can inhibit the subject's capacities to imagine a positive future or mobilise pro-environmental actions. As an increasingly common psychological condition, climate change anxiety, for instance, has been officially recognised by various expert bodies, including the UK's Natural Environment Research Council (2021) and the American Psychological Association (2021); see also Benoit et al., 2022). In this chapter, the authors explore how climate anxiety marks the consciousness of a growing number of contemporary youth in a manner that fundamentally shapes how they see the world and their place within it. But, equally, how a loss of the sense of a secure, safe future can be an important motivator to engage in actions to address such loss. Collective action, for instance, is thought to offer an important outlet for channelling fear, frustration, anger, and anxiety about the possibility of a 'non future' (Beck, 2015), or what Skotnicki and Nielsen (2021) refer to as the debilitating impacts of 'futurelessness', where prospects for realising a better world prove evermore difficult to imagine. Drawing on the findings of their own qualitative research, the authors explore youth imagination with respect to climate futures, noting how the participants (aged 18–30) in their study articulated anxiety about anticipated risks but also felt motivated to engage in various forms of activism that prove beneficial for the relief of anxiety, building stronger psychological understandings of climate change as a 'proximal experience' and awareness of how one can be an active agent of change, including restoring positive imaginative relationships with endangered futures.

In Chapter8, Sperati, Spinelli, Fasolo and Lionetti delve further into the diversity of experiences of climate change, calling for a nuanced approach to individual difference, as well as groups and cultures. Nature contact and our relationship with nature are fundamental to resilience-building. However, depending on a multitude of individual, physical and socio-cultural factors, people may have different opportunities to get in contact with nature and a different level of connectedness with the natural environment. As Setti, Stitzel, Ottoboni & McCarthy note in Chapter 6, neurodiversity is an example of such differences. Highly sensitive individuals tend to process environmental stimuli deeply, making them more vulnerable to degradation but also more empathetic and connected to nature in middle and older age. This connection enhances their pro-environmental attitudes. This chapter argues that high sensitivity can be leveraged to build resilience, reduce negative emotions linked to environmental crises, and empower people to respond actively to climate challenges. Specifically, it highlights the importance of early exposure to nature, particularly for highly sensitive children. Early-life exposure to green spaces is especially critical, as it can strengthen resilience in sensitive individuals, inspire environmental stewardship, and empower them to cope with the psychological impacts of climate change. This approach not only helps individuals build personal resilience but also encourages collective action to address environmental issues.

In the third section of this volume, contributors assess the experiences of Indigenous communities of climate change loss in various regions of the Arctic. As the United Nations Forum on Indigenous Peoples and Climate Change (2022) pointed out recently, Indigenous communities worldwide are amongst the most vulnerable to the effects of climate change (IPCC, 2022; United Nations, 2007) with those residing in rapidly changing Arctic regions particularly vulnerable (see, also, The Arctic Institute, 2023). In **Chapter9, Toivanen** highlights the role Indigenous peoples worldwide have played in protecting large portions of the world's biodiverse territories. Indigenous knowledge and practices of managing water, forests and ecosystems, developed over centuries of caring for nature, have slowed rates of deforestation and mining and in that, have played a vital role in the stewardship of nature, one potentially of enormous benefit to wider communities and practice of environmental care.

In spite of their pivotal role politically, the SDGs have not offered Indigenous peoples a comprehensive protection of their rights to date (e.g. UN Declaration on the Rights of Indigenous Peoples, 2007). Most notably, their right to self-determination (Article 4), the right to preserve one's cultural heritage (Article 5), one's land (Article 10), participate in decision-making on matters that affect them (Article 13) and be free from 'any kind of discrimination' (Article 2). All of these rights, as Toivanen observes, have been marginalised in global sustainable development frameworks, including the SDGs. Instead, priority is given to a modern scientific perspective on resilience-building or technological solutions to sustainability challenges where the contribution of Indigenous ecological knowledge systems and practices is deliberately sidelined, as are histories of systemic abuse and colonial exploitation. By highlighting the importance of Indigenous knowledge systems and capacities to overcome adversity, Toivanen showcases the value of their contribution to an epistemological, cultural and social change in approach to climate change mitigation and adaptation.

Similarly, in **Chapter10, Vladimirova** points to the residual capacities of Indigenous communities in the Russian Arctic to cope with disturbance, including the ever more destructive impacts of climate change. Vladimirova describes how Indigenous specialised knowledge of reindeer husbandry proves indispensable to the survival of resilient resource management systems and the safeguarding of local biodiversity for the future. In their examination of the experiences of the Forest Sami communities in the northern regions of Sweden, **Ohlsson and Lindh in Chapter11** likewise assess the opportunities available to these peoples to remain resilience. They note a consistent non-recognition of Forest

Sami peoples' rights to land, self-determination and agency in the realisation of sustainable Sami futures, is supported by a 'logic of elimination' where, over time, ownership of forest land and stewardship of remaining biodiversity is steadily transferred from Indigenous stewardship to state or commercial control. To date, an intense state-back industrial forestry initiative has seen 90% of Swedish forests turned over to industrial use. As a result, Sami peoples confront a continuous struggle to have infringements of their rights (e.g. the right to sustain traditional Indigenous ways of life and livelihood, the right to non-interference with Sami reindeer herding practices, the right to avail of local pasturelands, and consultation on new development initiatives, etc.) recognised legally and politically as undermining locally grounded resilience-building efforts. Particularly problematic is the way resilience continues to be interpreted in official policy discourse in a manner not conducive to the survival of Indigenous cultures but, rather, to expanding economic output at all—an important point of reflection given the call to reimagine ways of living sustainably, including in the Global North.

The virtues of Indigenous ways of life, in terms of their contribution to environmental rejuvenation and human resilience-building, are deliberately ignored. Indigenous varieties of resilience-building nurture spaces of synthesis and active renewal (Tuck & McKenzie, 2015) beyond experiences of destruction and marginalisation. The process of healing starts from the inside out through an 'active repudiation of dominance, tragedy and victimry' (see Vizenor, 1994) and the restoration of 'a native sense of presence', spirit and pride (Vizenor et al., 2014). Indigenous survivance offers guidance as to how to respond to climate-related loss (e.g. 'climate-ready' tribe's initiatives) and address the social, health and ecological risks global climate change poses to intergenerational connection. In spite of their value, Indigenous beliefs and practices of resilience building are actively overshadowed by the dominance of a 'science as solution' model of resilience in official sustainability policy discourse.

As local indigenous knowledge of the Arctic's land, sea and biodiversity plays a crucial role in the provision of essential services, including search and rescue (i.e. SAR) and the protection of the Arctic's natural surroundings, ensuring the continuity of this knowledge and Indigenous values more generally, as **Wood-Donnelly notes in Chapter 12**, is essential for ongoing resilience efforts. Stories of rescue and survival preserve a belief in the capacities of Indigenous communities to survive what are increasingly hazardous conditions created by rising temperatures and global climate change. Indigenous lives and everyday knowledge of safe travel routes, the predictability of weather patterns and hunting seasons is threatened by the effects of global interference with the climate system. As the 'life force' of the Arctic, snow, ice and permafrost ensure the continuity of the natural and social vitality of the Circumpolar North. Their disappearance would mark the end of its communities and rich heritage. Arctic resilience, therefore, requires the active support of its peoples and wider nature, including the right of both to enjoy the benefits of the cold (see the Inuit Petition, 2005).

In their critical appraisal of resilience-building efforts, **Sakshi and McCauley in Chapter 13** examine ongoing exploitations of Indigenous peoples' lands and other key resources as harmful practices that make resilience strategies necessary in the first place. Yet, the expectation is that Indigenous peoples will embrace resilience building programmes that are often at odds with traditional Indigenous philosophies of nature, programmes that define the current ecological crisis as a 'singularly environmental' concern with no regard for its various political, economic and socio-cultural dimensions. The authors offer further empirical evidence of these problematic tendencies and their contribution to enduring inequalities (e.g. forms of neo- colonialism, corporate resource grabbing, the ecological and social consequences of expanding mineral extraction industries in the Arctic) focusing on two case studies—Indigenous communities in the Northern Territory of Australia and in various regions of the Arctic, including the Canadian Arctic and the Republic of Sakha in Russia. In both instances, resilience measures designed to decrease the likelihood of disaster under highly unpredictable and rapidly changing environmental, social and geo-political conditions, are examined through a critical lens. While the Canadian approach to resilience is considered to fare better than the Russian one in terms of engagement with the concerns of Indigenous communities on the development of the Polar Code, the further commercial advancement of the Northern Sea Route, etc., prioritising the protection of the economic and geo-political interests of larger players in the Arctic, often at the expense of those of Indigenous communities in the region.

External pressures to embrace new commercial projects in the interests of nurturing the 'viability' of the wider regional economy are met with resistance from many Indigenous communities, including Inuit community representatives. Increasingly, these actors utilise digital platforms to promote their perspective on these issues and challenge the injustice dimensions of currently dominant narratives on what climate resilience or sustainable development requires? Similarly, Ohlsson & Lindh in Chapter 9 observe a growing willingness and propensity amongst Swedish Sami communities to legally challenge state renditions of a 'green transition' and its costs to traditional Sami reindeer herding practices through the courts in defence of their rights to continue to practice Indigenous models of resilience as survivance.

In the fourth section of this volume, authors examine the impacts of climate change on farming communities in various regions of Ireland and Iran, offering original field research on farmers' experiences of climate-related loss (e.g. loss of income, security of employment, farm heritage, identity, faith in a bright future, etc.), noting how these further aggravate underlining social, economic and health inequalities. Wider international research on the impacts of climate change has identified farming communities across world regions as a high-risk cohort (see Acharibasam, 2018; Cianconi et al., 2020; Tschakert et al., 2017) on account of their dependency on stable environmental conditions together with strong historical, cultural, and psycho-emotional connections to the land and related biodiversity. While the farming sector, especially largescale farming, has played a non-negligible role in accelerating rates of climate change, farmers also play a pivotal role in promoting biodiversity, and overall sustainability and have faced considerable challenges in adapting to changing environmental and social conditions.

Contributions to this section examine how multiple stress variables (deteriorating soil quality, crop yields, temperature rises, or the risk of flooding, drought, etc.) combine in ways that undermine basic capabilities, including the ability to enjoy a stable source of income, security of being, place, and identity and infringe on subjects' ontological security and basic human rights. Add to these problems the absence of an effective, regionally adaptable climate mitigation strategy that further undermines many farmer's commitments to governments' green agenda. Often described as the 'hidden costs' of climate change (Intergovernmental Panel on Climate Change, Working Group II, 2014), these challenges make the inclusion of the perspectives of farming communities especially pertinent. Contributors to this section explore these issues in depth, focusing on how farming communities most at risk in regions of Ireland and the Middle East interpret experiences of adversity and respond resourcefully to them.

In **Chapter14, Isaloo** examines the case of Lake Urmia, once one of the world's largest natural sea salt lakes located in Northwest Iran. In the last decades, Lake Urmia has progressively decreased to a tenth of its former size due to the cumulative effects of rising temperatures, intense drought, as well as largescale commercial developments in the region. The effects of these changes to date have been dire, especially on farmlands located in the river's basin region where more than 8% of Iran's food is produced. Isaloo notes how poor policy choices on water management and half-hearted efforts to overcome the detrimental effects of drought on the region have further undermined the resilience capacities of local communities (approximately 6.4 million people), leading to a steady increase in the numbers displaced ear year. Isaloo calls for a greater inclusion of local farming communities in resilience-building initiatives, particularly those whose knowledge of the land, local ecosystems and water conservation is extensive, to ensure a successful sustainable development strategy for the region is implemented. The absence of local farmers' voice is also noted by **McGurn in Chapter15**, where traditional High Nature Value farmland practices in the upland grasslands of the Aran Islands, Inis Mór, Inis Meáin, Inis Oírr, or the Burren on the west coast of Ireland are explored in terms of their contribution to protecting local biodiversity. McGurn highlights the absence of an appreciation for the contribution of these farming practices to climate resilience-building in government policy on climate change mitigation. Any new national legislation to promote carbon efficiency and meet legally binding emissions reduction targets (e.g. Ireland's Climate Action Plan, 2024) and biodiversity recovery plans must engage seriously with already established sustainability practices developed over centuries by local communities operating in local contexts. In other words, governments should not assume traditional farming practices are always contrary to sustainable development goals. To date, factors relating to social and cultural heritage have been neglected, as McGurn notes, including the vital role grazing animals play in maintaining low intensity agricultural systems and relatedly, plant species diversity of major conservation concern in the upland grasslands of the west coast of Ireland—areas of immense value to Ireland's wider socio-cultural, ecological and historical heritage.

Preserving established means of protecting biodiversity proves all the more pertinent when one considers, as **McNamara & O'Dwyer do in Chapter16** the type of climate changes anticipated for Ireland over the next decades, with winter rainfall projected to increase by upwards of 20%, further exacerbating the risk of flooding and more intense and frequent storms, particularly in coastal and low-lying regions. Such changes are particularly detrimental to Irish agricultural practices, freshwater conservation and maintaining infrastructural resilience. However, as the worst impacts of these changes are still expected to be felt in the future, opportunities still exist, as McNamara & O'Dwyer note, to develop more effective adaptation and mitigation approaches that work with farmers on the ground. The authors point to the need to further develop and support progressive initiatives such as the Signpost Programme, a whole sector approach to supporting Irish farmers in their transition to a more sustainable future. This programme consists of a network of demonstration farmers who engage in a series of training programmes designed to facilitate intense farmer-to-farmer learning (e.g. workshops) with one-on-one target support services leading to farm-specific action plans being devised for each farmer participating in the programme. As McNamara & O'Dwyer observe, resilience-building is a key aim of the Signpost Programme, as is fostering greater trust between sustainability advisors and farming communities as to what practices, technologies and support services (e.g. AgNav digital platform) will best facilitate farmers' transition to or, indeed, maintenance, of green and healthy farming practices. The adaptive capacities of farming communities to adjust to climate-induced changes reflect not only farmers' material and physical abilities to adapt to changing conditions and new societal expectations, but, equally, their socio-cultural political and psycho-emotional capacities to adapt to changing conditions as well. What is essential, therefore, as the authors note, is that farmers feel their voice is heard and importantly, they remain part of an inclusive and non-threatening transition process moving forward.

In the concluding section of this volume, the editors reflect on its major findings, noting the significance of observed commonalities and differences in the experience of climate impacts amongst various world communities and societal groups, and how these observations might inform a more holistic, democratic approach to resilience-building (that addresses the environmental, social, mental, physical, and civic health of communities). That is, a model of resilience capable of fostering hope, wellbeing and the recovery of democratic commitments to cooperative actions informed by the possibilities and potentials (as much as the limits and dangers) of the current democratic age to protect, manage and share planetary resources peacefully and effectively into the future.

Notes

1. The European Council defines an 'economy of wellbeing' as one that fosters a governance and policy approach that places people and their wellbeing at the center of decision-making and understands 'wellbeing' as an integral component of 'development' (for further discussion, see European Council's Conclusions on the Economy of Wellbeing, 2019; Dyar et al., 2022).
2. See Schutte, Torquati and Beattie (2017), Ferreira et al. (2006).

References

American Psychological Association. (2021). *Addressing climate change concerns in practice*. Available at: https://www.apa.org/monitor/2021/03/ce-climate-change.

American Psychological Association (2023). *Sustainability, climate and the natural environment*. Available at: https://www.apa.org/topics/climate-change#:~:text=What%20APA%20is%20ng&text=APA%20Services%20works%20to%20support,promote%20adaptation%20to%20its%20effects.&text=2023%20report%20from%20APA%20and,children%20and%20youth%20mental%20health.

Australia Council of Social Service (ACOSS). (2015). *Resilient community organisations. Disasters, climate change and disadvantage*. Available at: https://resilience.acoss.org.au/the-six-steps/introduction/disasters-climate-change-and-disadvantage.

Beck, U. (2015). Emancipatory catastrophism: What does it mean to climate change and risk society? *Current Sociology, 63*(1), 75–88.

Benoit, L., Thomas, L., & Martin, A. (2022). Review: Ecological awareness, anxiety, and actions among youth and their parents - a qualitative study of newspaper narratives. *Child and Adolescent Mental Health, 27*(1), 47–58.

Bronfenbrenner, U. (1979). *The ecology of human development*. Harvard University Press.

Cianconi, P., Betrò, S., & Janiri, L. (2020, March 6). The impact of climate change on mental health: A systematic descriptive review. *Frontiers in Psychiatry, 11*, 74. https://doi.org/10.3389/fpsyt.2020.00074. PMID: 32210846; PMCID: PMC7068211.

Dyar, O. J., Haglund, B. J. A., Melder, C., et al. (2022). Rainbows over the world's public health: Determinants of health models in the past, present, and future. *Scandinavian Journal of Public Health, 50*(7), 1047–1058. https://doi.org/10.1177/14034948221113147

European Commission. (2022). *European Commission announces Dublin and Cork among 100 cities participating in EU Mission for climate-neutral and smart cities by 2030*. Available at: https://ireland.representation.ec.europa.eu/news-and-events/news/dublin-and-cork-among-100-cities-participating-eu-mission-climate-neutral-and-smart-cities-2030-2022-04-28_en

European Council (2019). *Council conclusions on the Economy of Wellbeing (2019/C 400/09)*. Available at: https://eur-lex.europa.eu/legal-content/EN/TXT/PDF/?uri=CELEX:52019XG1126(06)&rid=5#:~:text=The%20Economy%20of%20Wellbeing%20is%20a%20policy%20orientation%20and%20governance,of%20wellbeing%20and%20economic%20growth

Ferreira, K., Van der Horst, W., Wendel-Vos, S., Kremrs, F. J. V., & Lenthe, J. B. (2006). Environmental correlates of physical activity in youth—A review and update. *Obesity Reviews, 8*(2), 129–154.

Intergovernmental Panel on I Climate Change, Working Group II. (2014). *Impacts, adaptation and vulnerability*. Available at: https://www.ipcc.ch/report/ar5/wg2/

Intergovernmental Panel on Climate Change (2022). *IPCC sixth assessment report: impacts, adaptation and vulnerability*. Available at: https://www.ipcc.ch/report/ar6/wg2/.

Inuit Circumpolar Conference (2005). *Petition to the Inter American Commission on human rights seeking relief from violations resulting from global warming caused by acts and omissions of the United States*. Available at: https://www.ciel.org/Publications/ICC_Petition_7Dec05.pdf

Irish Government Climate Action Plan (2024). Available at: https://www.gov.ie/en/publication/79659-climate-action-plan-2024/

Oregon Health Authority (OHA) (2023). *Climate and Health in Oregon 2023*. Available at: https://www.oregon.gov/oha/PH/HEALTHYENVIRONMENTS/CLIMATECHANGE/Documents/FINAL%20Climate%20Health%20in%20Oregon%202023%20v071124%20(1).pdf

Lerner, S. (2010). *Sacrifice zones: The front lines of toxic chemical exposure in the United States*. MIT Press.

Schutte, A. R., Torquati, J. C., & Beattie, H. L. (2017). Impact of urban nature on executive functioning in early and middle childhood. *Environment and Behaviour, 49*(1), 3–30.

Skillington, T. (2023). Where does the true value of a frame analysis lie. *Discourse Studies, 25*(2), 288–296.

Skotnicki, T., & Nielson, K. (2021). Toward a theory of alienation: Futurelessness in financial capitalism. *Theory and Society, 50*(6), 1–29.

Soga, M., & Gaston, K. (2016). Extinction of experience: The loss of human-nature interactions. *Frontiers in Ecology and the Environment, 14*(2), 94–101.

The Lancet (2020). *The 2020 report of the lancet countdown on health and climate change*. Available at: https://www.thelancet.com/journals/lancet/article/PIIS0140-6736(20)32290-X/abstract

Tschakert, P. Barnett, J. Nevill, EE, Lawrence, C, Tuana, N. New, M., Elrick-Barr, C., Pandit, R., & Pannell, D. (2017). Climate change and loss, as if people mattered: values, places, and experiences. *WIREs Climate Change, 8*(5). https://doi.org/10.1002/wcc.476.

Tuck, E., & McKenzie, M. (2015). Indigenous methods and critical place inquiry. In M. McKinley (Ed.), *Place in research: Theory, methodology, and methods* (pp. 129–142). Routledge.

UK Natural Environment Research Council (2021). *Programme launched to tackle climate anxiety in young people*. Available at: https://www.ukri.org/?st=anxiety

United Nations Declaration on the Rights of Indigenous Peoples (2007). Available at: https://chr-observatories.uwazi.io/en/entity/cqf2kyyvc38/toc?gad_source=1&gclid=Cj0KCQiA2oW-BhC2ARIsADSIAWp-eUTFejQ63rSQOHJrShJ6BtZDlHHsVr1OfW-0CEjxQ0473hcjMqIaAsXCEALw_wcB&page=5

United Nations Environment Programme (UNEP) (2024). *Smarter and inclusive cities*. Available at: https://www.undp.org/sites/g/files/zskgke326/files/2024-06/smarterandinclusivecitiescourse_2.pdf.

Vizenor, G. R. (1994). *Manifest manners: Post Indian warriors of survivance*. Wesleyan University Press.

Vizenor, G., Tuck, E., & Yang, K. W. (2014). Resistance in the blood. In E. Tuck & K. W. Yang (Eds.), *Youth resistance research and theories of change*. Routledge.

World Health Organization (2025). *Climate change and health: Shaping a resilient future*. Available at: https://www.who.int/europe/news-room/events/item/2025/02/28/default-calendar/climate-change-and-health-shaping-a-resilient-future#:~:text=28%20February%202025%2C%2010%3A00,heatwaves%20and%20wildfires%20each%20year

Part I

The Climate Resilient City

2

Acknowledging Homelessness in the Climate Resilient City: Sanctioning Homelessness and Enabling Environmental Risk

Stephen Przybylinski

Introduction

Climate resilience has been integral to sustainability and just transition discourses within urban geography and planning scholarship for decades. Studies abound that examine how urban ecosystems are affected by, and can adapt to, changing climatic conditions (Tyler & Moench, 2012), as well as those addressing how urban governance equitably deliberates and implements environmental and climate policies (Boyd & Juhola, 2015). Less prevalent within this broad scholarship are critical approaches examining the effects of urban governance on an expanding group who is especially vulnerable to climate impacts: the unhoused. While initiatives seeking to foster climate resilient cities through sustainable development goals regularly acknowledge socio-economic indicators as being central to

S. Przybylinski (✉)
Geography, Environment, and Spatial Sciences, Michigan State University, East Lansing, MI, USA
e-mail: przybyli@msu.edu

T. Skillington and A. Setti (eds.), *Climate Change Resilience Across Societal Contexts*, Energy, Climate and the Environment,
https://doi.org/10.1007/978-3-031-95020-9_2

these efforts (e.g., UNDP, 2023), rarely is homelessness explicitly considered within them (Casey & Stazen, 2021). Indeed, the City of Portland, Oregon, U.S.A, the city from which this chapter's case emanates, makes no mention of homelessness within its own "climate emergency workplan," despite the concept of resilience remaining prominent throughout its pages (City of Portland, 2022). While municipal decision-making on urban homelessness is regularly uncoupled from urban climate adaptation planning, this chapter seeks to illustrate the value in bringing them together, by detailing the case of homelessness encampment development in a "pioneer" city in sustainable development planning.

From the late-twentieth century into the early twenty-first, the City of Portland was touted for being at the vanguard of sustainable development approaches (Goodling et al., 2015; Slavin & Snyder, 2011; Smith, 2014). More recently, however, it has become known for its nontraditional approaches to managing a rising homeless population. Since declaring a state of emergency on housing and homelessness in 2015, for instance, the city government worked to institutionalize a network of self- and non-profit-managed homeless encampments as an alternative to traditional congregate shelter. By April 2021, the city had legalized sanctioned encampments and tiny house villages by incorporating them into its city code as official land use types. As of 2024, there are 16 encampments and tiny house villages that range in size from 10 to 75 people per site (in addition to two mass encampments). Though not without some of the same problems as congregate shelters, these forms of outdoor emergency shelter are critical not just for saving lives, but significantly, and for self-managed encampments in particular, providing residents a means of directly controlling their own living environment while they are housing unstable. Research on self-managed encampments shows that these sites enable residents the agency to make decisions about everyday tasks and to participate in working together as a community to protect their own interests (Przybylinski, 2021a, 2022; Sparks, 2017, 2024). Bandaging the wounds of a larger housing crisis, peer- and non-profit-managed encampments offer temporary relief from the rigor of living without housing in urban America.

While there are clear social and political benefits to this model of emergency shelter, due to the speed at which these services are developed, in addition to the limited available properties on which they can operate, outdoor encampments are often sited on locations not meant for human habitation. As a result, many encampments expose residents to elevated environmental and climate harm. The most common is extreme heat. Portland's climate sits somewhere between a warm and hot summer Mediterranean climate, where winters are cool and rainy and summers are hot and dry. And summer peak heat there is getting dangerous. From 2020 to 2023, Portland experienced 80 days with temperatures between 90- and 99-degrees Fahrenheit (32.2–37.2 Celsius), and 16 days with temperatures at or above 100 degrees Fahrenheit (37.7 Celsius). A major heat wave in June 2021 saw three respective days with highs of 108 (42.2 C), 112 (44.4 C), and 116 (47 C) degrees Fahrenheit (Friesen, 2024). During this particular "heat dome" event, four unhoused Portlanders died due to overheating (Walters et al., 2023, p. 6). While the intensity and duration of extreme heat poses unequal risks across race and geography (Hoffman et al., 2020), other environmental effects, like air and industrial pollution, present further environmental risks for unhoused communities as well.

Detailing research from self-managed encampments in Portland, this chapter describes how residents at two encampments experience environmental and climate risks and how this impacts the well-being of residents in these communities. In doing so, the chapter extends insights from critical environmental justice scholarship to suggest that environmental and climate injustices must also be examined through intersecting injustices. While the presence of sanctioned homeless encampments within the urban landscape clearly reflects economic inequities within society, connecting environmental and climate injustices to encampment resident experiences illuminates how routine environmental harms go unnoticed as critical factors that exacerbate social inequities. I conclude by arguing that we should not disentangle socio-economic struggles of unhoused communities from the environmental inequities degrading their lived experiences and that any policy pathway toward more stable housing conditions requires the reduction of avoidable environmental and climate risks. The failure of not only Portland's government, but all municipal governments, to address such inequalities will merely exacerbate the effects of climate change for the most vulnerable.

Relating Homelessness to Environmental and Climate Justice

The breadth of environmental justice (EJ) and climate justice (CJ) scholarship is enormous. Decades of research centered on EJ and CJ has produced an array of studies spanning disciplines and themes. It is surprising then that EJ and CJ research has for the most part avoided examining the injustices related to environmental and climate harms that unhoused people in particular experience. This is not to say that there is *no* research exploring how environment and climate affect unhoused people, however. A nascent literature exists addressing how the well-being of unhoused communities is affected by environmental and climate changes.

In general, EJ and CJ scholarship has traditionally assessed the distributive, procedural, and recognitional forms of justice (Schlosberg, 2007; Walker, 2009). EJ scholars, for instance, examine where and to what extent environmental burdens or benefits are differently distributed among specific groups (Bullard, 1983), as well as how democratic or inclusive participation is concerning environmental decision-making, and who may be recognized as a subject of justice in these discourses (Bell & Carrick, 2017; Fraser, 1997). Similarly, CJ scholars have focused on the iniquitous outcomes of changing climate through distributive, procedural, and retributive/recognitional forms of justice (see Skillington, 2023). And while EJ and CJ scholarship has expanded upon the forms through which it assesses and explains injustices (Ohlsson & Przybylinski, 2023), these traditional forms framing environmental and climate injustice guide much of this scholarship today.

These traditional forms—distributive, procedural, and recognitional—largely shape what little research exists on homelessness as it relates to environmental and/or climate justice as well. Such studies often examine distributive inequities among housed and unhoused populations. For example, one study examining distribution and access to publicly available potable water in Athens, Greece, Istanbul, Turkey, and Los Angeles, U.S.A., three cities with historically large homeless populations, found access to publicly available water severely lacking. "What urban inhabitants experiencing homelessness have in common across these three cities," the author notes, "is the basic need for water in a cityscape warmed by rising temperatures and increased aridity" (Hale, 2019, p. 36). To lessen environmental inequity, Hale argues that cities ought to increase the number and availability of publicly accessibly water fountains to ensure life-saving measures for those most vulnerable to water insecurity. A similar finding from a case study of Phoenix, Arizona, USA, reveals that unhoused people living in encampments are differently affected by water insecurity compared with other forms of being unhoused. They note that for some encampments, "isolation from [water] resources [makes] water access more likely to be unconventional and less safe" (DeMyers et al., 2017, p. 80). For instance, the study finds that encampment communities draw water from sources such as "flood water, rainwater, canal water, and water that collects in retention zones" (p. 76). And while most of this accessed water was used for bathing or cooling down, they also found that surface water and private taps were used for drinking in dire circumstances (p. 76). Along with distribution and access to water, other studies have examined the effects of air pollution on unhoused people. One recent study of homelessness in Salt Lake City, Utah, USA, found that 89% of their participants sought medical attention "because of a condition associated with poor air quality," and further, that there was no statistically significant difference between negative air quality health outcomes for chronic and nonchronic unhoused people (DeMarco et al., 2020, p. 10). Together, these studies illustrate the intimate relationship between environment/climate and one's experience being unhoused, by underscoring how differences in living circumstances between housed and unhoused populations are further challenged by environmentally related inequities. It is these types of issues that will likely persist without direct incorporation into climate action plans.

Homelessness scholarship concerned with assessing how and to what extent environmental and climate injustices affect the well-being of unhoused people, therefore, follow from the premise that changing climate conditions and environmental risks put unhoused people at more risk than they normally would be regarding safety and health. Indeed, the U.S. federal government acknowledges the reality that extreme weather events exacerbated by changing climate (e.g., floods and wildfires) drives people into homelessness (USICH, 2022). And much research globally has examined the uneven processes of resettlement and shelter stability due to climate-driven migration (Kidd et al., 2021). Yet, less scholarship has focused on examining how the circumstances of being unhoused are relationally affected by climate change/environmental risks. Unhoused people regularly experience risks to climate and environment that housed people are often able to mitigate. But as Goodling (2020, p. 835) rightly argues (following Pellow's [2018] approach to critical environmental injustice), a critical analysis of homelessness through EJ or CJ ought to require "attention not only to direct exposure to hazards, but also to criminalization of houseless people." For, when unhoused communities like encampments experience environmental harms, for instance, municipalities regularly "respond to hazard exposure with displacement, creating a cycle criminalization, [and thus] further exposure to environmental hazards" (p. 835). Goodling highlights a contradiction in relying upon the state to mitigate environmental harms, whereby regulating encampments due to elevated risk of environmental harms often displaces unhoused people from one toxic site to another. And it is for this reason, Goodling argues, that criminalization [of the unhoused], serial eviction, and exposure to environmental hazards go hand-in-hand" (p. 851). In this way, the cyclical relationship between the state and unhoused communities becomes a central point of concern for critically understanding environmental/climate injustices for the unhoused.

Goodling was primarily referring to unsanctioned, but nevertheless, organized encampments, and the effects that living in toxic spaces have on these individuals, especially when accounting for racial and gendered differences among these communities. Unorganized encampments are particularly susceptible to being disbanded by police or local government. But even Portland's self-managed encampments, communities that have been operating on the same site for years, and therefore, are relatively safe from displacement *because* they are sanctioned, face a similar problem. Many are sited in unsafe environments where, regardless of being permitted to operate there, can be displaced with municipal authority if complaints are made. This, of course, is an issue not only for the unhoused people being exposed to different harms, but for municipalities that are supposed to uphold environmental standards for living environments. Yet, if Portland were to disband all sanctioned encampments to the scale that the city now hosts, it would displace nearly 500 people back into similarly vulnerable spaces. The rest of the chapter describes how unhoused people living in two sanctioned encampments in Portland are exposed to environmental/climate harms before addressing how these issues must also be seen as social inequities concerning not just their potential displacement, but their emplacement in spaces with safer living standards.

Climate Risks in Portland's Homeless Encampments

In 2023, Portland counted its highest number of unhoused people in its history, 6297, a number that had been increasing on average for more than a decade (Multnomah County, 2023). Of the total unhoused population, its unsheltered population specifically increased 29% from the previous year alone, making "unsheltered" homelessness the majority condition for unhoused people in the city (ibid). Particularly concerning with these changes is that there has been a significant rise in the number of unhoused people of color in the city relative to the general population. For unsheltered people specifically, the number of Black, Indigenous and other people of color (BIPOC) counted in the 2022 census increased by 21% since the previous count three years prior, along with a 7% increase of unsheltered non-Hispanic whites. Notably, people identifying as American Indian, Alaska Native or Indigenous made up 12.9% of Portland's unhoused population, despite making up only 2.5% of the County's general population. People identifying as Black or African American made up 17.2% of the unhoused population, despite making up 7.1% of the Multnomah County population (Elliot & Oschwald, 2022). And while no official numbers exist detailing the number of residents living in the city's network of self-managed and non-profit managed encampments, the capacity set for each encampment in the city collective sits somewhere between 400–500 people who are living in outdoor shelters throughout the city.

Dignity Village (DV) and Right 2 Dream Too (R2DT) are Portland's two longest operating encampments, DV since 2004 and R2DT since 2011. Both are self-managed encampments, meaning that they are contracted by the city to manage their own membership and operations, though they are sited on municipally owned property. The two encampments model the democratic ethos that has inspired the more recent spate of encampments and villages throughout the city, and they are resilient social spaces operating through the sweat equity of members who ensure that these alternatives to congregate shelter continue to function. The locations of the two encampments nevertheless expose their residents to certain environmental conditions that affect their residents' well-being.

DV began in the early 2000s as a roving protest encampment in downtown Portland. After a year-long struggle to operate its own encampment, the city government permitted DV to operate on a plot of government owned property roughly 7 miles (11 km) north of Portland's downtown (Przybylinski, 2024). The site was created by partitioning off a quarter lot of a municipal composting facility, a still active site bounding DV to its west. To its immediate south is a low-level correctional facility, and approximately one and half football fields to the east lies a runway to Portland's International Airport. The encampment infrastructure sits on asphalt and has no surrounding trees to block the consistent summer sun. The space is comprised of about 60 tiny homes constructed with various materials that were built mostly by residents over its twenty-year history, in addition to a few common buildings and a complex of non-flushable toilets and showers. While many residents feel safer there than on the street or in congregate shelter, its tiny homes provide modest reprieve from the elements and do not feature prominently within planning discourses around climate resilience (Fig. 2.1).

Some elements of risk for DV residents reflect those that also face housed people. Take extreme summer heat, for example. Due to how the individual tiny houses maintain heat, and given the lack of air conditioning in them, DV residents struggle to stay cool during the day and

Fig. 2.1 Dignity Village commons buildings, with municipal compost facility in background

sometimes at night. For these reasons, residents convene their weekly membership and council meetings outside in the shade of the commons building during meetings. But being outside at DV has its tradeoffs. For instance, during weekly meetings, residents frequently need to pause conversation because of the noise from jets taking off from the nearby runway. At one meeting at DV in 2018, I counted 15 take-offs flying over DV within an hour, a process requiring residents to cease talking for about a minute per take-off. While noise pollution is one annoyance, air pollution is another. Though no air quality measures exist for DV itself, studies of the effects on residents and workers living and working in proximity to airports routinely show increased levels of ultrafine particulate matter (UFP) from commercial aircraft. One review of the literature measuring airline impacts on air quality in nearby areas found consistently elevated levels of UFPs outdoors as well as indoors, with off-airport sites measured from as far away as 12 km (Riley et al., 2021, p. 7). Of particular concern with UFPs is the elevated levels of nitrogen oxides and black carbon for areas off-site, but nearby airport runways (Ibid, p. 7). The relative dearth of air quality data from airports, and the lack of standard measurement and instrumentation to capture UFPs, makes it difficult to identify the direct health impacts on adjacent residences. But studies have shown slight decreases in cardiopulmonary and respiratory (e.g., diminished lung function) outcomes for people 300 meters away from a commercial airline runway (Riley et al., 2021, p. 8, citing Lammers et al., 2020). The site of DV sits between 900–1000 meters from the runway.

Though the extent to which DV residents have respiratory issues due to the barrage of airplanes flying overhead is uncertain, residents there consistently complained about respiratory issues during my interviews. Some figured it was likely not good practice to live next to an airport runway, but did not know how that affected them specifically. Overwhelmingly, however, longer term residents complained particularly about the respiratory impacts of the municipal composting facility bordering the encampment site. Some residents did not know that the municipal facility was recycling organic waste piles, though more than one person witnessed employees wearing hazmat suits while working on the waste piles. They noted that the processing of the yard waste would send layers of brown and black dust into the camp. As one resident told me, "we noticed that when you wash your plate, set it down, ready for a barbecue and then you swipe your finger across it, there would be a black smear [on it]." For that reason, they said, "we also never left [out] a glass of liquid to drink because you may have a layer of stuff floating on top of it." Understanding that this dust was just part of everyday existence throughout the encampment, one resident noted that his recently-washed clothes would be dirty after hanging them out to dry. "So, I try to wash on their [compost workers'] off days." "You still don't wear white around here because everything turns grayish, brownish," he said. It was likely these dust particles, residents say, that affected their respiratory symptoms.

One DV resident noted that, shortly after moving in, she began to have trouble breathing. "Within six months of being here, I was on one inhaler," she said. "Within a year they deemed me as a severe asthmatic … had me on another inhaler. Because, walking from our commons to my home [200 steps away], I was having to hit my inhaler two to three times." She noted that things have since gotten slightly better for her after medication changes, but nevertheless noted that at one point shortly after her arrival, "everyone was on an inhaler." Two other DV residents spoke similarly about how the encampment's location next to the compost facility triggered their respiratory issues. One male resident complained that he got "migraine headaches from it. My nose runs constantly. And even with the antihistamines, it still runs constantly." Yet, he said, "if I go leave here for an hour, no more runny nose." All three residents interviewed above noted that these types of respiratory irritations were common among others at the encampment as well.

When I asked DV residents if they were worried about their health, one noted the irony of the site's services and location. They said:

> I try not to think about it. I mean, we're in a situation that we're stuck in. And I mean, I'd love to see my way out of it, but I can't. I'm stuck right now. And that's the way it is. And quite frankly, I need this place. And this is what I got to do. But the fact is ... I would think, because of what's coming up over that fence, and when I'm breathing in, [it] is probably causing something similar to black lung, or something of that nature. Because, I have respiratory problems anyways. I smoke a lot. And, quite frankly, I have asthma. And probably COPD. I would assume that the fibers are causing some form of black lung, or something of that nature. Because it's got to be getting embedded inside your lungs.

For these and other residents of DV, dealing with constant respiratory problems is simply part of the cost of living without traditional housing. Few options exist for them to move into safer shelter, away from where particles from organic waste flow freely into living spaces, where noise and air pollution disrupt speech and sleep, and where severe heat renders comfort during summer months hard to come by. Given that the decision of siting encampments is nearly always out of the hands of the unhoused communities living there (though see Przybylinski, 2021b), residents' well-being because of these forms of environmental risk are outweighed by the immediacy of simply having *some* place to sleep.

R2DT

R2DT began similarly to DV, as a protest encampment in downtown that eventually was permitted by the city to operate near the city center on municipally owned land. R2DT has about 16 members who live in wooden tiny houses, and the members themselves run a day and night shelter on site. It is the city's first low-barrier shelter, where overnight sleepers rest on the floor of a large tent. Weekly meetings there are held in these tents, and in the summer it can be oppressively hot, and in the winter, shivering cold. These are the sleeping environs for about 70 guests there each night (Fig. 2.2).

While R2DT is located much closer to resources in the urban core than DV, its location is also squarely within industrial space. Wedged between a grain silo that loads wheat from eastern Oregon and ships to the rest of the world, a cargo and passenger rail line going north–south from Seattle to San Diego, and one of the city's light rail lines, the encampment is constantly inundated with noise and air pollution. When the grain elevator was operating, R2DT residents complained of "snowing" grain that would fall over the encampment. But this inconvenience

Fig. 2.2 R2DT encampment seen from afar, centered among grain tower and cargo and light rail lines

was minor compared to the regular issues that the abutting industries posed for the site. On March 18, 2018, an hour before R2DT was to hold their weekly meeting, a cargo train derailed less than 20 meters away from the encampment. 6 cars derailed, one of which was a tanker normally holding toxic chemicals. Fortunately, the tanker was empty, as the train was returning to Portland from outside the city. But this stretch of track has seen repeated derailments around this exact corner and is known to the Union Pacific railway company as a place of concern. The residents at the encampment were a bit shaken by the event, but as many noted to me, there were not really any other options of where the encampment could be sited. This was hand they were dealt (Fig. 2.3).

Other toxic threats abut R2DT as well. Sitting on the grounds of the now defunct grain elevator site is a rubber tire recycling "center." Though there really is no center to speak of, the company uses the land in between the train tracks and the silo to store rubber pellets from chopped-up tires which are then shipped to Asia. On May 25, 2023, a massive fire started on a large pile of rubber pellets there, sending acrid smoke in all directions, including into R2DT. The fire was contained but reignited four days later due to residual heat. It should go without saying that the toxins from incinerated rubber are harmful for humans

Fig. 2.3 Derailed train car abutting the R2DT encampment

to ingest. Nevertheless, the fact that large piles of rubber exist (with no regulation) adjacent to an outdoor encampment, it bears repeating which toxins emerge from these repeated fires. Incinerating rubber is incredibly harmful to humans' respiratory and cardiovascular systems, as well as vital organs. Emissions from burning rubber contain a range of toxins, including benzene and chloroform, and heavy metals such as lead, chromium, and zinc (Center for Health, Environment and Justice, 2011, p. 7). Dioxins and furans are also released, which are dangerous to the central nervous system (ibid). And though the Oregon Department of Environmental Quality fined the company responsible and forced them to obtain a permit to store open-air rubber (Effinger, 2023), this official action merely permits rubber to accumulate uncontained, enabling future fire events to occur that affect not only R2DT, but the surrounding soil and water nearby.

While other environmental and climatic risks exist for residents in both of these encampments, such as the lack of access to clean water at a third encampment where I worked (Hazelnut Grove), central to these encampments' issues relates to their location. The process of siting encampments is a difficult one. As one city planner I spoke with who worked on the city's more recent "safe rest villages" indicated, the number of bureaucratic criteria that restrict where temporary accommodations like encampments can be is numerous. It is hard enough, they noted, to find a property that has electric hookups with some access to water, let alone one that accords with fire and other building codes for tiny houses. Indeed, the city is currently conducting an environmental cleanup on a brownfield site for a new mass encampment (Wozniacka, 2023). And as the city continues to develop more and larger homeless encampments, it seems likely that the only properties available will be on former industrial sites.

Conclusion

This chapter focused on two self-managed homeless encampments in one liberal-progressive city. Though the specific environmental and climate harms experienced by residents there are not generalizable to all unhoused encampments, a few lessons from these residents' experiences are nevertheless worth noting. The first emanates from the very process of selecting sites to sanction for emergency shelter. If and when municipalities decide to permit organized encampments, frequently the chosen sites are located within industrial zones that would not otherwise be amenable to human habitation. The fact that an encampment operates outdoors means that its residents will likely be exposed to more harms than those living in traditional housing. While a few of Portland's newer encampments are tucked into more traditional neighborhood spaces, where they may occupy commercial or small residential lots, the conscious decision to site encampments on industrial sites underscores that encampment residents' exposure to environmental and climate harm is at once facilitated by the state while enforcing residents' immobility in these sites. As such, residents must accept that where they live exposes them to specific environmental harm. For, they risk being displaced should they complain about ill health because of onsite source pollutants and extreme heat. Despite being an emergency service, and thus making residency in these sites short term, encampments should not be Excepted from common environmental standards. Organized outdoor shelters must have protections just like those living in traditional housing commonly enjoy and one means toward doing so would be to incorporate these types of standards within climate initiatives for municipal governments, like Portland's own climate emergency work plan.

A second stems from the nascent homelessness scholarship on EJ and CJ. Despite an abundance of scholarship detailing unhoused peoples' lived experiences in general, little research connects unhoused peoples' experiences with direct environmental harms and the effects of changing climatic conditions. Even less has advanced a critical EJ or CJ approach to examining homelessness. A critical EJ or CJ lens invites us to analyze how it is that certain groups of people are more subject to the circuitous processes of criminalization, displacement, and eventual emplacements where exposure to environmental harms is high. Given that initiatives like the United Nation's sustainable development goals suggest that socio-economic indicators ought to be encompassed in climate resiliency planning (UNDP, 2023), such an approach also provides a lens through which to assess how one aspect of social inequity heightens risk for environmental and climate inequity. In Portland, like many other U.S. cities, Black and Indigenous peoples are disproportionately experiencing higher rates of homelessness relative to the general population. While no official numbers exist detailing demographic information of individuals residing in the city's 16 encampments today, my fieldwork in sanctioned and self-managed indicates that the identities of residents in these communities generally reflect the demographic composition of the unhoused population throughout the city. Thus, to the extent that unhoused people of color are living in self-managed encampments sanctioned by the city, we ought not disentangle the relationality of environmental and climate harms with socio-economic inequities that residents of color, queer or transgendered people, or disabled people experience before, during, and after housing instability.

Such insights sit uneasily with the continued sanctioning of encampments across the U.S.A, where state-led initiatives to house vulnerable populations often reinforce the marginality of unhoused people. At the same time, the relative permanence of temporary shelter communities allows us to critically reflect on these types of citing decisions for encampments to better understand how the social circumstances of sanctioned encampment communities make residents distinctly vulnerable to environmental and climate harm. As such, any policy pathway intended to establish more stable housing conditions for unhoused communities requires the reduction of avoidable environmental risks, a recognition that must be sensitive to why certain groups of people are more impacted by the socio-spatial cycle of homelessness that enables further environmental and climate risk.

References

Bell, D., & Carrick, J. (2017). Procedural environmental justice. In *The Routledge handbook of environmental justice* (pp. 101–112). Routledge.

Boyd, E., & Juhola, S. (2015). Adaptive climate change governance for urban resilience. *Urban Studies, 52*(7), 1234–1264.

Bullard, R. (1983). Solid waste sites and the black Houston community. *Sociological Inquiry, 53*, 273–288.

Casey, L., & Stazen, L. (2021). Seeing homelessness through the sustainable development goals. *European Journal of Homelessness, 15*(3), 63–71.

Center for Health, Environment and Justice. (2011). *Burning rubber: Tire incineration.* Falls Church, Virginia. [PDF]. Available at: https://chej.org/wp-content/uploads/Burning%20Rubber-%20PUB%20101.pdf

City of Portland. (2022). The city of Portland's 2022–2025 climate emergency workplan. Available at: https://www.portland.gov/bps/climate-action/documents/climate-emergency-workplan-2022-2025/download

DeMarco, A., Hardenbrook, R., Rose, J., & Mendoza, D. (2020). Air Pollution-related health impacts on individuals experiencing homelessness: Environmental justice and health vulnerability in Salt Lake County, Utah. *International Journal of Environmental Research and Public Health, 17*(22), 8413. https://doi.org/10.3390/ijerph17228413

DeMyers, C., Warpinski, C., & Wutich, A. (2017). Urban water insecurity: A case study of homelessness in Phoenix, Arizona. *Environmental Justice, 10*(3), 72–80.

Effinger, A. (2023, August 7). Owner of Combustible Tire Pile on the Willamette Logs Another Blaze at Shredding Plant in North Portland. *The Willamette Week*. Available at: https://www.wweek.com/news/2023/08/07/owner-of-combustible-tire-pile-on-the-willamette-logs-another-blaze-at-shredding-plant-in-north-portland/

Elliot, D., & Oschwald, M. (2022). *2022 point-in-time methodology report. Count of Homelessness in Portland, Gresham, Multnomah County, Oregon.* Joint Office of Homeless Services. https://www.multco.us/multnomah-county/news/full-2022-point-time-count-report-shows-covid-19-added-unsheltered

Fraser, N. (1997). *Justice interruptus: Critical reflections on the "postsocialist" condition.* Routledge.

Friesen, M. (2024). Portland temps, 1938–2024. *The Oregonian.* Available at: https://projects.oregonlive.com/weather/temps/

Goodling, E. (2020). Intersecting hazards, intersectional identities: A baseline critical environmental justice analysis of US homelessness. *Environment and Planning E: Nature and Space, 3*(3), 833–856.

Goodling, E., Green, J., & McClintock, N. (2015). Uneven development of the sustainable city: Shifting capital in Portland Oregon. *Urban Geography, 36*(4), 504–527.

Hale, M. (2019). Fountains for environmental justice: Public water, homelessness, and migration in the face of global environmental change. *Environmental Justice, 12*(2), 33–40.

Hoffman, J. S., Shandas, V., & Pendleton, N. (2020). The effects of historical housing policies on resident exposure to intra-urban heat: A study of 108 US urban areas. *Climate, 8*(1), 12.

Kidd, S. A., Hajat, S., Bezgrebelna, M., & McKenzie, K. (2021). The climate change–homelessness nexus. *The Lancet, 397*(10286), 1693–1694.

Multnomah County (2023). *Chronic homelessness number falls across tri-county region in 2023 Point in Time Count.* Available at: https://www.multco.us/multnomah-county/news/news-release-chronic-homelessness-number-falls-across-tri-county-region-2023#:~:text=The%20full%20Point%20in%20Time,disparities%20and%20demographics%20in%20homelessness.&text=Multnomah%20County%20reported%20a%2016,to%202%2C610%20counted%20in%202023

Ohlsson, J. and S. Przybylinski, eds. (2023). *Theorising justice: A primer for social scientists.* Bristol University Press.

Pellow, D. (2018). *What is critical environmental justice?* Polity Press.

Przybylinski, S. (2024). From rejection to legitimation: Governing the emergence of organized homeless encampments. *Urban Affairs Review, 60*(1), 118–148.

Przybylinski, S. (2022). Without a right to remain: Property's limits on Portland's self-governing houseless encampments. *Environment and Planning C: Politics and Space, 40*(8), 1711–1226.

Przybylinski, S. (2021a). Realizing citizenship in property: Houseless encampments and the limits of liberalism's promise. *Political Geography, 91*, Article 102494.

Przybylinski, S. (2021b). Securing legal rights to place: Mobilizing around moral claims for a houseless rest space in Portland, Oregon. *Urban Geography, 42*(4), 417–438.

Riley, K., Cook, R., Carr, E., & Manning, B. (2021). A systematic review of the impact of commercial aircraft activity on air quality near airports. *City and Environment Interactions, 11*, Article 100066.

Schlosberg, D. (2007). *Defining environmental justice: Theories, movements and nature.* Oxford University Press.

Skillington, T. (2023). Climate Justice. In *Theorising justice: A primer for social scientists.* Bristol University Press. Available at: https://doi.org/10.51952/9781529232233.ch010

Slavin, M., & Snyder, K. (2011). Strategic climate action planning in Portland. In M. Slavin (Ed.), *Sustainability in America's cities: Creating the green metropolis* (pp. 21–40). Island Press.

Smith, R. M. (2014). Planning for urban sustainability: The geography of LEED®–Neighborhood DevelopmentTM (LEED®–NDTM) projects in the United States. *International Journal of Urban Sustainable Development, 7*(1), 15–32.

Sparks, T. (2024). *Tent city, Seattle. Refusing homeless and making a home.* University of Washington Press.

Sparks, T. (2017). Citizens without property: Informality and political agency in a Seattle, Washington homeless encampment. *Environment and Planning A, 49*(1), 86–103.

Tyler, S., & Moench, M. (2012). A framework for urban climate resilience. *Climate and Development, 4*(4), 311–326.

UNDP (United Nations Development Programme). (2023). *Building climate-resilient and inclusive cities.* Available at: https://www.undp.org/sites/g/files/zskgke326/files/2023-12/policy_brief-building_climate-resilient_and_inclusive_cities.pdf

USICH (United States Interagency Council on Homelessness). (2022). *Help the Biden-Harris administration tackle the triple threats facing America: climate change, homelessness, and housing shortages.* Available at: https://www.usich.gov/news-events/news/help-biden-harris-administration-tackle-triple-threats-facing-america-climate

Walker, G. (2009). Beyond distribution and proximity: Exploring the multiple spatialities of environmental justice. *Antipode, 41*(4), 614–636.

Walters, J., Evans, B., Vines, J., and Lewis, P. (2023). *Domicile unknown: Review of deaths among people experiencing homelessness in Multnomah County in 2021.* Multnomah County, Oregon. Available at: https://multco-web7-psh-files-usw2.s3-us-west-2.amazonaws.com/s3fs-public/Domicile-Unknown-2021.pdf

Wozniacka, G. (2023, December 13). Industrial contamination at proposed Portland mass shelter site raises alarm. *The Oregonian.* Available at: https://www.oregonlive.com/environment/2023/12/industrial-contamination-at-proposed-portland-mass-shelter-site-raises-alarm.html

3

A Peoples-Centred Approach to the Climate Resilient City: Addressing Inequalities in Access to Nature and Its Restorative Benefits

Tracey Skillington and Johanna Marie Kirsch

Introduction

In 2022, the European Commission selected Cork, the second largest city in the Republic of Ireland, as one of one hundred 'exemplary' European cities committed to becoming a climate-neutral by 2030. Inspired by international commitments to improve cities' connectivity with nature, Cork City Council has pledged to become a 'trailblazer' in the realization of the EU's climate-smart agenda.[1] However, the city faces certain challenges in this regard, not least an acceleration in rates of loss of native plants, local habitats and animal species (The National Biodiversity Data Centre, 2023), an expanding population (which has grown

T. Skillington (✉)
Department of Sociology & Criminology, University College Cork, Cork, Ireland
e-mail: t.skillington@ucc.ie

J. M. Kirsch
Cork, Ireland

T. Skillington and A. Setti (eds.), *Climate Change Resilience Across Societal Contexts*, Energy, Climate and the Environment,
https://doi.org/10.1007/978-3-031-95020-9_3

annually by 1% since 2006, especially in suburban regions), rising crime rates, heightened competition for land and other essential resources—trends that largely mirror those occurring in other cities across the world today (Colding et al., 2020).

The following discussion reports on the findings of a qualitative study conducted between January and December 2023 exploring commitments to the rejuvenation of the city's biodiversity and eco-cultural heritage, noting the benefits to mental, physical and social wellbeing of regular engagement with its green and blue spaces, as well as the impacts of restricted access to the same for those with limited opportunities to avail of these resources.[2] The research suggests a partial membership of the city's sustainable development initiatives is granted to those nominally included in government's 'transformative climate action plans' (European Commission, 2024) but effectively excluded in practice due to a limited recognition of their needs. This has the effect of reinforcing certain socially embedded patterns of inequality that map onto wider systems of age, gender, race, socio-economic and disability injustice that need to be addressed as a matter of urgency, given their impact on the lives of many. A series of proposals are presented to address the distributive, recognitional, procedural, and epistemic dimensions of overlapping injustice and improve prospectives for enacting a more inclusive model of sustainable development for the city.

The 15-Minute City as a Remedy for Interconnected Challenges?

In 2016, Carlos Moreno and his research team introduced the concept of 'the 15-minute city' (Moreno, 2016). At its core is the idea that the urban dweller should be able to meet all their basic daily needs within a 15-minute bike ride or walk from their home with essential services, including access to green spaces, in close proximity. So began the campaign to promote the 15-minute city as the epitome of sustainable living, reduced traffic congestion and improved quality of life for all the city's inhabitants. According to Moreno, the 15-minute city serves as a 'regeneration' model to promote the rights of all to health, irrespective of place of residence, socio-economic status, age, or mobility capacity (Moreno et al., 2021). In the years since its introduction, the concept of the 15-minute city has gained considerable momentum in sustainable urban planning initiatives with cities such as Paris, Barcelona, Melbourne, Portland, and Vancouver promoted as important examples of the 15-minute city.

Reviewing plans to further transform the city of Oslo in Norway into a 15-minute city, Akrami et al., (2024) assess the practical viability of this model, focusing on the Hovinbyen area of the city. They found that, apart from the central part of Oslo, many of the wider suburban regions of Oslo do not fit easily into the 15-minute model and scored low on the issue of accessibility. Considering these findings, in addition to the steady rise in the population density of Oslo's suburbs, the authors concluded the 15-minute city model did not, in fact, offer a valid 'regeneration model' for many of Oslo's urban areas but, rather, presented a useful tool to assist city planners in the design of new urban areas.

For its critics (Weng et al., 2019), the 15-minute city is wrongly presented as a 'one size fits all' model that frequently fails to balance the city's need for conservation and the protection of cultural heritage with convenience imperatives (Khavarian-Garmsir et al., 2023). Equally, understandings of the varying capacities of urban residents to walk to amenities, depending on the age and physical health of the individual or, indeed, safety of their surroundings, is not taken sufficiently into consideration in applications of this model and sustainable urban redesign more generally (Willberg et al., 2023). By highlighting the sociological relevance of these issues, the research findings documented below draws attention to some of the ways in which the city's sustainable development project can limit the accessibility of essential resources for some in local settings (O'Mahony et al., 1996). The attainment of the UN's sustainable development goals continues to face serious challenges. One perhaps less well explored challenge is the restricted access of many of the city's residents to its green and blue spaces and with that, limited opportunities to avail of their health-restoring benefits.

As conservation behaviours and values have been shown to depend heavily on regular engagements with nature, this predicament could potentially pose a serious challenge to efforts to halt or reverse current alarming rates of biodiversity loss, as well as commitments to sustainable development (Gaston & Soga, 2020). Over the last two decades, psychologists have gathered increasing evidence of a reduction in urban populations' familiarity with acoustic (e.g. knowledge of different types of bird song, the sounds of rivers, waterfalls, etc.) and aromatic dimensions of local biodiversity (e.g. the scent of native flowers, trees, and other plant life) and their life-enriching qualities. This decline is largely attributed to decreasing levels of exposure to nature. So advanced are these developments, Soga and Gaston (2018) see their onset as signalling an 'extinction' of experiences of nature. Viewing and engaging with natural worlds occurs increasingly today through flat screens or fast-moving vehicles (Marty-Dugas et al., 2018). Consequently, behaviours crucial to maintaining good mental, physical and social health, including daily physical activities (Melkevik et al., 2010), exposure to sunlight and fresh air (Schutte et al., 2017), as well as in-person social interaction (Twenge, et al., 2019) are said to be declining.

Given these developments, Smalley (2022) queries whether it will be possible to generate sufficient motivation to care about the demise of nature if large numbers of people no longer consciously register its presence in their daily lives? The research of Gaston and Soga (2020) points to the importance of routine exposure to the soundscapes, scent, and visual displays of woodlands, parks, rivers, trees, flowers, insects, birds, etc., to the public's motivation to act to protect biodiversity and address climate change. The question then is whether the ability to care about nature's wellbeing and the planet more generally is retrievable if time spent immersed in nature continues to decline in the future?

The research outlined below sought to test the assumptions of this literature and consider if a decline in access to and motivation to care for nature should be taken as given across all communities? In particular, the notion that the rise of sedentary lifestyles, a growing dependency on phones and other smart devices, increasing levels of obesity, and disconnection from natural settings are largely self-imposed limitations, or lifestyle choices, affecting our psycho-emotional relationships with nature. There is clear evidence to suggest that at least some of these are also shaped by factors linked to class, age, socio-economic background, environmental changes, and urban structural design (e.g. dense road networks, a shortage of public parks or sports facilities, safe walking routes, reliable transport services, etc.). For the mobility challenged, the visually impaired, the wheelchair-bound, the elderly, or young children, travelling to and from city parks, river walks, etc., can be a hazardous journey due to heavy city traffic, poor air quality, rising temperatures, antisocial behaviour, uneven or broken pathways, etc. (e.g. see Bromley et al., 2007). For these cohorts, access to nature's restorative health benefits may be less influenced by lifestyle choices than societally conditioned factors that determine how fast an 'extinction of the experience of nature' occurs and at what level of intensity.

If an 'extinction of the experience of nature' can be detected, it is most likely to affect some people more severely than others. Factors such as age, gender, mobility level, residential area, socio-economic status, and underlying health have been shown to work in tandem to negatively influence access to the restorative benefits of the city's green and blue spaces (Shanahan et al., 2014). The recent work of Hoffman et al., (2020, see also HelpAge International, 2016), for instance, shows how structurally embedded inequalities are reproduced in experiences of the effects of rising temperatures, for instance, on city life (e.g. the creation of heat islands, as well as an over-use of heat-retaining building materials in poorer residential areas, including asphalt and concrete). The question of who has easy access to nature and its restorative effects requires further investigation.

As Colley et al., (2022; see, also, World Bank, 2023) observe, climate lives are not lived equally but vary significantly across contexts and cohorts, locally and internationally. The effects of climate change and biodiversity loss are not only felt as geological changes (changes in the physical structures and substances of the Earth), but also as powerful limitations on freedom (e.g. freedom from want, freedom of movement). The capacity to adapt one's life conditions to a climate changing world and a sustainable redesign of one's city is heavily conditioned by social disparities in the availability of choice (Skillington, 1998, 2017, 2019). A growing body of research points to problems encountered by the elderly (HelpAge International, 2016), persons with a disability (World Bank,

2023) and youth from poorer socio-economic communities (Sanson et Bellemo, 2021) in availing of such choices in their daily lives. As this literature highlights, the needs of these cohorts are rarely prioritized in policies aimed at strengthening communities' climate resilience (see, also, Nissen et al., 2020).

To take fuller advantage of the health-replenishing effects of the city's green and blue spaces, as governments and health experts recommend (e.g. WHO, 2023), further investment is needed in a people-centred approach to climate resilience, one capable of measuring the impacts of climate change driven structural transformations of the city (changes in transport, energy, housing, healthcare, urban design, etc.) on the quality of life for all residents, and taking action to address any disadvantages encountered. A failure to resolve these issues may result in a situation where prevailing sustainable development policies and resource management procedures are encountered as discriminatory, especially by those with a limited capacity to avail of new services. By shifting the focus somewhat to the question of opportunity to engage with the city's green and blue spaces (in addition to the issue of availability), attention comes to be placed on the ways in which the city's sustainable development plans may restrict the choices available to some.

A social stratification of ecological choice reflects a 'patterning' of inequalities (Hoerning, 1971) across different issue domains. Limited.

opportunities to avail of the health and wellbeing-enhancing dimensions of the city's green and blue spaces thereby become yet another expression of enduring inequalities experienced by marginal groups. Such inequalities not only impact the physical health of those affected but overall wellbeing as well. Our research shows how limited access can intensify feelings of not belonging to one's city, where belonging is understood as not only a knowing the rules of eco-social engagement in one's community but also being involved in their creation (May, 2011). In highlighting these issues, we share with Nussbaum (2003) the view that factors affecting the implementation of a truly equitable development programme cannot be understood sufficiently without a detailed investigation of the welfare of its subjects.

Research Study

Our research sought to engage closely with city residents to explore these issues in live or immersive everyday social settings between January and December 2023. Supported by generous funding from the Irish Research Council's New Foundations Programme (2022), as well as the Sunflower Charitable Foundation through The Community Foundation for Ireland and in partnership with Cork Nature Network, our research was conducted in two phases. The first phase consisted of fifty walking interviews conducted with a representative sample of the city's population. Our objective here was to gain a better understanding of respondents' awareness of the potential risks declining environmental conditions and poor resource management practices pose to the future flourishing of the city's biodiversity, eco-cultural heritage and general wellbeing.

The findings of the walking interviews (n = 50) drew attention to the key importance of regular immersion in the city's green and blue spaces to mental health (94% of respondents reported that the primary reason for them visiting a city park regularly was to protect their mental health). Nature's cycles of growth, decay and rebirth were said to offer respondents reassurance that nature was doing fine despite wider climate change challenges and evidence of biodiversity loss. Witnessing plants grow and tree foliage bloom reportedly provided respondents with a sense of security, belonging and hope for the future. For those living in the inner city, being over-exposed to a heavily grey, concrete environment with little greenery was noted as a risk to their health and mental well-being. 48% of respondents referred to the therapeutic value of specific aspects of nature, including hearing the sounds of nature (varieties of birdsong, the babbling of streams and rivers, etc.), noting its contribution to mood enhancement and general wellbeing (see, also, Hartig et al., 2014). A further 22% of respondents recounted the pleasure they derive from the aroma of seasonal shrubs, woodlands, flowers, and grasses. The symbolism and positive associations evoked by the sounds, scent, physical features, and cycles of nature in the city's parks and blue spaces were repeatedly highlighted, particularly in their capacity to create a 'feeling of being away' from the noise and chaos of the city and allowing psychological recovery to occur (Kaplan, 1995).

> Spring in the park makes me feel like there is life all around us. Kind of hopeful, I guess. That's why it's nice to go for walks here because it makes you feel better. (Respondent 36, female, Irish born, Fitzgerald's Park, Cork City)
>
> … it's definitely interesting because the cycles of nature also kind of remind you that things will end and there will be new beginnings, you know. (Respondent 41, male, Irish born, Marina Park, Cork City)

Discussing such issues with respondents while walking in their world allowed for an exploration of the rich meanings people invest in the city's green and blue 'spaces' as 'places' regularly visited and as 'sites' where identities are forged, memories are made, wellbeing restored and new insights gained on nature's flourishing or, indeed, demise (Gagnon et al., 2015). Respondents were interviewed as they walked through one of eight city parks, all of which are under the jurisdiction of Cork City Council, a metropolitan district council responsible for the provision of housing, roads, public transportation, urban amenities, planning and environmental protection.[3] Respondents were asked to consider factors that potentially enhance or undermine their and fellow residents' access to nature in the city. Access here was defined as:

- Physical access to the city's green and blue spaces,
- Sensory access (the ability to hear, see, smell and feel a sensory awareness of nature's presence),
- Socio-cultural access to the city's nature (the ability to connect with a nature that forms a key part of the city's identity & history).

In particular, the discussion focused on how these various forms of access are facilitated or, indeed, hindered by external factors, such as urban design, the quality of transport links, green spaces in the inner city, the volume of traffic and anti-social behaviour, as well as environmental conditions.

In phase two of our research, focus groups were organized during the summer months of 2023 in four residential areas of the city to explore these issues further and consider what factors limited residents' access to the city's green and blue spaces and their restorative benefits. At this stage in our research, access to nature was explored as a relevant theme of distributive, recognition, procedural and epistemic justice, with justice understood as the intentional integration of knowledge, people, and possibility for the better of all planetary life, guided by a principle of non-domination (Bohman, 2008: 190).[4] Our research considered if opportunities to enjoy nature's restorative benefits were distributed evenly across various socio-economic communities? Second, if sufficient recognition was being granted to the needs and capacities of various cohorts within these communities?

Wellbeing not only requires that adequate resources are available to achieve basic human functioning but, also, that individuals have sufficient capacity to access them (Dyar et al., 2022; Nussbaum, 2003). In other words, nature must be accessible to all if individuals are to experience its benefits, physically, mentally, and culturally. Third, did focus group participants feel their concerns about the state of nature in the city were being heard (procedural justice) and further, if their contributions to recent plans to redesign the city along more sustainable lines were taken on board and reflected in new policy actions (i.e. epistemic justice)? Our aim was to better understand how these different types of social justice intersect in people's perceptions and lived experiences of their city's 'green transition'.

Focus group participants drew attention to what they described as serious shortfalls in efforts to address inequalities in access to the city's nature and decision-making processes around the same. Journeying to and from city parks and nature trails was seen as dangerous for those with physical, auditory, or visual impairments, or parents of young children, for instance. Issues such as antisocial behaviour, uneven footpaths, speeding traffic and dangerous walkways were highlighted as deterrents to walking or cycling to city parks and blue spaces. One respondent with a visual impairment noted:

> Last Wednesday, I can remember being really anxious to get home because you can just feel the negative energy on the roads. Especially in finer weather, drivers are more anxious to get home, they are rushing to get

> home to relax so they are stressing to get to a more relaxed space. (Focus Group 2, Participant 1)

Heavy traffic undermines the confidence of many of the city's more vulnerable residents who question the safety of their movements through the city when it is busy:

> ..they put cycle lanes all along my road. It's the same for Mary [next door neighbor]. We both have difficulties in walking so it's a major inconvenience for both of us. (Focus Group 1, Participant 1)

One of the positive aspects of the Covid 19 pandemic noted by respondents was the time it gave people with their city. Emptied of traffic, noise, and crowds, people could re-engage with the city's nature (mentioned by 11% of respondents) and benefit from its wellbeing-enhancing effects:

> It gave us the opportunity to admire the night sky again and reflect more on the wildlife around us. From my bedroom window in Montenotte, we could see the swallows gathering near the Marina every day. We all loved to watch their antics. (Focus Group 4, Participant 1)

These moments with the city were thought to have since largely dissipated as traffic congestion, pollution, and road reconstruction again mark everyday experiences of the city. Opportunities to enjoy its green and blue spaces are once more curtailed by a sense of being at risk for many. Focus group respondents noted that the city's new sustainable development initiatives (e.g. redirected traffic flows) disadvantage those whose needs are not taken sufficiently into consideration in the design of the 'climate-smart city' (e.g. the mobility challenged) with priority still being accorded to the needs of motorists and now e-scooter users also (65% of respondents). Poorly positioned cycle lanes, speeding motorists and more recently, e-scooter users, along with rising crime rates, were singled out as particularly problematic. All were thought to limit the safe access of many to the city's green and blue spaces and negatively impact socialization patterns and overall wellbeing.

Several focus group participants also highlighted disparities in the quality of park facilities available to residents on the north and south sides of the city. Historically, the north side of Cork City has had a larger number of working class and socially disadvantaged communities. Persisting inequalities in access to recreational and sports facilities on the north side of the city were noted by many focus group participants as a potential barrier to achieving the city's sustainable development goals (i.e. SDG 3 Good Health and Wellbeing; SDG 10 Reduced Inequalities and SDG 11 Sustainable Cities and Communities) (raised in discussion in focus groups one and three):

> Compared to facilities, and the level of maintenance of parks on the Southside, we are very much the poor relation. The south side gets everything. They have more parks, better services. (Focus Group 3, Participant 5)

The failure of local government to address these issues appropriately was seen to further exacerbate existing inequalities, especially when the voice of poorer socio-economic communities is repeatedly ignored:

> ...we were just spoken down to and left feeling like rubbish really at the end of it [public meeting to discuss transportation redesign plan for the city]. How much money did they spend on that? You know doing this lip service thing where they are spending all this money but not actually doing anything to help lift people out of poverty and improve their quality of life. (Focus Group 1, Participant 1)

Protecting access to nature is a necessary precondition for the enjoyment of many rights, including the right to health (UNDHR, 1948, Article 25) and free movement (ibid., Article 13) and impacts a broad range of additional rights, including security of person (physical and mental) (Article 3), free association (Charter of Fundamental Rights of the European Union, 2012, Article 12), the rights of the elderly (ibid., Article 25), and the integration of persons with disabilities (ibid, Article 26). Disrupted access to nature, its health-restoring and rights protecting benefits, therefore, may be seen as actively compromising the basis upon which justice can be performed effectively. Stronger efforts are needed to address current limitations on access to nature as a fundamental justice issue.

Conclusion

A growing body of research points to the multiple benefits derived from regular contact with the city's green and blue spaces, noting how they contribute to the building of more resilient and healthy communities (e.g. see Addas, 2023; Barboza et al., 2021; European Environment Agency, 2022). While we broadly agree with the argument of this research, we note how efforts to create sustainable and inclusive cities today are often hampered by a failure to address underlining inequalities. The most salient of these, as noted by the respondents in our research, are those relating to unreliable public transport links, unsafe pathways, heavy traffic, inadequate support structures (seating, canopy, etc.) and security risks (due to rising crime and anti-social behaviour in the city) that disproportionately affect the aged, women, the disabled, youth and poorer socio-economic communities. Arguably, our findings also support those of Hassell and the Irish institutional Property Group (IIP) (2021) who recently investigated the ten most important services Irish publics would like access to within fifteen- minutes of their home. Of all age cohorts, interviewees aged 55 or above listed easy access to public transport (57%) and green and blue spaces (56%) as priority concerns. What our research findings and those of the of Hassell and the IIP Group suggest is that while there is a strong desire amongst older cohorts to spend more time in the city's green and blue spaces and enjoy their restorative benefits, opportunities to do so can be limited by various external factors (including inadequate public transport services, the risk of robbery, insufficient funds, etc.), excessive traffic and pollution levels dangerous to health. As identified limitations on access to nature were largely externally sourced rather than linked purely to lifestyle choice, attitudes to exercise, etc., there is, we argue, a notable stratification of immersive experiences of the city's green initiatives, one that affects vulnerable cohorts more readily than others. That said, we do acknowledge that to be able to illustrate the limiting power of this stratified system of access to nature in the city and its negative effects in more detail, a more extensive study of its institutional mechanisms is needed, one that is beyond the scope of this chapter.

However, this research does highlight a need to pay more attention to. these issues as ones insufficiently explored in current official representations of the climate-smart city (European Commission, 2023) or the '15-minute city' model. Arguably, these omissions represent a blind spot in current policies aimed at promoting inclusive sustainable cities (UNEP, 2021). More careful consideration ought to be given to the ways in which plans to build climate-smart cities can, if insufficiently inclusive of the basic needs and relatedly, rights of vulnerable members of the community, undermine commitments to a just green transition. To ensure all perspectives are accounted for, focus group participants suggested that local government representatives be brought on walking tours of the city with residents with mobility issues, special needs (e.g. visual or physical impairments), young children or low income families, etc., to ensure that they witness first-hand the type of challenges such people face when attempting to enjoy the city's green and blue spaces and devise, on that basis, a sustainable development plan that is truly contextually grounded, 'immersive' and focused on addressing current barriers in the way of improving all peoples' access to green and blue spaces in the city.

Responding proactively to these challenges is essential if city councils are to foster a truly transparent democratic process that benefits municipalities by strengthening local support for the implementation of new sustainable development initiatives (Nilsson et al., 2021). Arguably, this requires a reappraisal of current SDG implementation and assessment measures to ensure 'climate-smart' city re-development is centred on informing, consulting, involving, collaborating with and empowering all people in their capacities to access and benefit from nature's life-enhancing potentials.

Notes

1. For example, the EU's new Nature Restoration Law (2024) which sets a binding target for the EU to restore 20% of its land and sea areas by 2030. Available from: https://environment.ec.europa.eu/topics/nature-and-biodiversity/nature-restoration-law_en, accessed April 16, 2024.
2. For a more detailed account of the findings of this research, see Skillington and Kirsch (2024).
3. The eight parks chosen for this research are the most regularly frequented in the city. Cork City Council (2024) Services. Available at: https://www.corkcity.ie/en/council-services/services/. Fifty interview respondents were recruited via social media, including the webpages of various residents' association and Cork Nature Network, as well as local community hall notice boards.
4. The political ideal of non-domination is defended most notably in neo-republican political thinking where it is thought to embody a conception of political freedom that serves as a foundational element of state citizenship.

References

Addas, A. (2023). The importance of urban green spaces in the development of smart cities. *Front Environmental Science, 11*, 1206372. https://doi.org/10.3389/fenvs.2023.1206372

Akrami, M., Sliwa, M. W., & Rynning, M. K. (2024). Walk further and access more! Exploring the 15-minute city concept in Oslo, Norway. *Journal of Urban Mobility, 5*, Article 100077. https://doi.org/10.1016/j.urbmob.2024.100077

Barboza, E. P., Cirach, M., Khomenko, S., et al. (2021). Green space and mortality in European cities: A health impact assessment study. *Lancet Planet Health, 5*, e718–e730. https://doi.org/10.1016/S2542-5196(21)00229-1

Bohman, J. (2008). Nondomination and transnational democracy. In C. Laborde & J. Maynor (Eds.), *Republicanism ad political theory* (pp. 159–216). Wiley-Blackwell.

Bromley, R. D. F., Matthews, D. L., & Thomas, C. J. (2007). City centre accessibility for wheelchair users: The consumer perspective and the planning implications. *Cities, 24*, 229–241. https://doi.org/10.1016/j.cities.2007.01.009

Colding, J. & Gren, A. & Barthel, S. (2020). The incremental demise of urban green spaces. *Land, 9*(162). https://doi.org/10.3390/land9050162

Colley, K., Irvine, K. N., & Currie, M. (2022). Who benefits from nature? A quantitative intersectional perspective on inequalities in contact with nature and the gender gap outdoors. *Landscape and Urban Planning, 223*, Article 104420. https://doi.org/10.1016/j.landurbplan.2022.104420

Cork City Council (2024). *Services*. https://www.corkcity.ie/en/council-services/services/.

Dyar, O. J., Haglund, B. J. A., Melder, C., et al. (2022). Rainbows over the world's public health: Determinants of health models in the past, present and future. *Scandinavian Journal of Public Health, 50*, 1047–1058. https://doi.org/10.1177/14034948221113147

European Commission (2023). *EU Mission: Climate-neutral and smart cities*. https://www.google.com/search?q=EC+cimate+smart+100+cities&oq=EC+cimate+smart+100+cities&aqs=chrome..69i57j69i64.6457j0j15&sourceid=chrome&ie=UTF-8

European Commission (2024). *EU missions to address climate change in cities and regions*. https://op.europa.eu/en/publication-detail/-/publication/45c6c574-3366-11ef-b441-01aa75ed71a1/language-en.

European Environment Agency (2022). Who benefits from nature in cities? Social inequalities in access to urban green and blue space across Europe. https://www.eea.europa.eu/downloads/803a2dd9755941439d2f70d5fa1d2e19/1675774183/who-benefits-from-nature-in.pdf

European Union Nature Restoration Law (2024). Available from: https://environment.ec.europa.eu/topics/nature-and-biodiversity/nature-restoration-law_en.

European Union Charter of Fundamental Rights of the European Union (2012). 2012/C 326/02, https://eur-lex.europa.eu/legal-content/EN/TXT/?uri=CELEX:12012P/TXT

Gagnon, M., Jacob, J. D., & McCabe, J. (2015). Locating the qualitative interview: Reflecting on space and place in nursing research. *Journal of Research in Nursing, 20*, 203–215. https://doi.org/10.1177/1744987114536571

Gaston, K. J., & Soga, M. (2020). Extinction of experience: The need to be more specific. *People Nat, 2*, 575–581. https://doi.org/10.1002/pan3.10118

Hartig, T., Mitchell, R., De Vries, S., et al. (2014). Nature and health. *Annual Review of Public Health, 35*, 207–228. https://doi.org/10.1146/annurev-publhealth-032013-182443

Hassell and the Irish Institutional Property Group (2021). *Close to home—exploring 15-minute urban living in Ireland (and beyond).* htps://www.hassellstudio.com/research/close-to-home-exploring-the-15-minute-urban-living-in-ireland-and-beyond.

HelpAge International (2016). *Making urban spaces work for older people.* https://www.helpage.org/silo/files/ageing-and-the-city-making-urban-spaces-work-for-older-people.pdf.

Hoerning, K. H. (1971). Power and social stratification. *The Sociological Quarterly, 12*, 3–14. https://doi.org/10.1111/j.1533-8525.1971.tb02085.x

Hoffman, J. S., Shandas, V., & Pendleton, N. (2020). The effects of historical housing policies on resident exposure to intra-urban heat: A study of 108 US urban areas. *Climate, 8*(12). https://doi.org/10.3390/cli8010012

Kaplan, S. (1995). The restorative benefits of nature: Toward an integrative framework. *Journal of Environmental Psychology, 15*, 169–182. https://doi.org/10.1016/0272-4944(95)90001-2

Khavarian-Garmsir, A. R., Sharifi, A., Hajian Hossein Abadi, M., et al. (2023). From garden city to 15-minute city: A historical perspective and critical assessment. *Land, 12*, 512. https://doi.org/10.3390/land12020512

Marty-Dugas, J., Ralph, B. C. W., Oakman, J. M., et al. (2018). The relation between smartphone use and everyday inattention. *Psychol Conscious, 5*, 46–62. https://doi.org/10.1037/cns0000131

May, V. (2011). Self, belonging and social change. *Sociology, 45*, 363–378. https://doi.org/10.1177/0038038511399624

Melkevik, O, Torsheim, T., Iannotti, R.J., et al. (2010). Is spending time in screen-based sedentary behaviours associated with less physical activity: A cross national investigation. *International Journal of Behavioral Nutrition and Physical Activity, 7*(46). https://doi.org/10.1186/1479-5868-7-46

Moreno, C. (2016). *La ville du quart d'heure: Pour un nouveau chrono-urbanisme.* https://www.latribune.fr/regions/smart-cities/la-tribune-de-carlos-moreno/la-ville-du-quart-d-heure-pour-un-nouveau-chrono-urbanisme-604358.html.

Moreno, C., Allam, Z., Chabaud, D., et al. (2021). Introducing the "15-Minute City": Sustainability, resilience and place identity in future post-pandemic cities. *Smart Cities, 4*, 93–111. https://doi.org/10.3390/smartcities4010006

Nilsson, A., Vanhuyse, F., Arra, V., et al. (2021). *Citizen participation for sustainable cities*. https://www.sei.org/wp-content/uploads/2021/05/factsheet-citizen-participation-for-sustainable-cities.pdf.

Nissen, S., Prendergast, K., Aoyagi, M., et al. (2020). Young people and environmental affordances in urban sustainable development: Insights into transport and green and public space in seven cities. *Sustainable Earth, 3*, 1–12. https://doi.org/10.1186/s42055-020-00039-w

Nussbaum, M. (2003). Capabilities as fundamental entitlements: Sen and social justice. *Feminist Economics, 9*(2–5), 33–59. https://doi.org/10.1080/1354570022000077926

O'Mahony, P., & Skillington, T. (1996). Sustainable development as an organizing principle for discursive democracy. *Sustainable Development, 4*, 42–51. https://doi.org/10.1002/(SICI)1099-1719(199603)4:13.3.CO;2-5

Sanson, A., & Bellemo, M. (2021, August). Children and youth in the climate crisis. *BJPsych Buletin, 45*(4), 205–209. https://doi.org/10.1192/bjb.2021.16.

Schutte, A. R., Torquati, J. C., & Beattie, H. L. (2017). Impact of urban nature on executive functioning in early and middle childhood. *Environment and Behaviour, 49*, 3–30. https://doi.org/10.1177/0013916515603095

Shanahan, D. F., Lin, B. B., Gaston, K. J., et al. (2014). Socio-economic inequalities in access to nature on public and private lands: A case study from Brisbane, Australia. *Landscape and Urban Planning, 130*, 14–23. https://doi.org/10.1016/j.landurbplan.2014.06.005

Skillington, T., & Kirsch, J. M. (2024). Assessing inequalities in access to the city's green and blue spaces through the experience of its residents. *Urban Resilience and Sustainability, 2*(3), 272–288.

Skillington, T. (2019). *Climate change and intergenerational justice*. Routledge.

Skillington, T. (2017). *Climate justice and human rights*. Palgrave.

Skillington, T. (1998). The city as text: Constructing Dublin's identity through discourse on transportation and urban re-development in the Press. *British Journal of Sociology, 49*(3), 456–473.

Smalley, A. (2022). Forest 404: Using a BBC drama series to explore the impact of nature's changing soundscapes on human wellbeing and behavior. *Global Environmental Change, 74*, Article 102497. https://doi.org/10.1016/j.gloenvcha.2022.102497

Soga, M., & Gaston, K. J. (2018). Shifting baseline syndrome: Causes, consequences and implications. *Frontiers in Ecology and the Environment, 16*, 222–230. https://doi.org/10.1002/fee.1794

The National Biodiversity Data Centre (2023). https://biodiversityireland.ie/ipbes-irelands-biodiversity-crisis/. Accessed April 16, 2024.

Twenge, J. M., Spitzberg, B. H., & Campbell, W. K. (2019). Less in-person social interaction with peers among US adolescents in the 21st century and links to loneliness. *Journal of Social and Personal Relationships, 36*, 1892–1913. https://doi.org/10.1177/0265407519836170

United Nations (1948). *Universal declaration of human rights*. https://www.ohchr.org/sites/default/files/UDHR/Documents/UDHR_Translations/eng.pdf

UN Environment Programme (2021). *Equitable future cities hold answers to pollution, climate and nature breakdown*. https://www.unep.org/news-and-stories/press-release/equitable-future-cities-hold-answers-pollution-climate-and-nature.

Weng, M., Ding, N., Li, J., et al. (2019). The 15-minute walkable neighbourhoods: Measurement, social inequalities and implications for building healthy communities in urban China. *Journal of Transport & Health, 13*, 259–273. https://doi.org/10.1016/j.jth.2019.05.005

Willberg, E., Fink, C., & Toivonen, T. (2023). The 15-minute city for all?—Measuring individual and temporal variations in walking accessibility. *Journal of Transport Geography, 106*, Article 103521. https://doi.org/10.1016/j.jtrangeo.2022.103521

World Bank Group (2023). *Disability inclusion*. https://www.worldbank.org/en/topic/disability.

World Health Organization (2023). *How can we assess the benefits of urban nature*. https://www.who.int/europe/news/item/22-05-2023-how-can-we-assess-the-benefits-of-urban-nature.

4

Building Resilience and Synergies for Healthy Air Practices Through Community Engaged Research: Learnings from the BEHAV-I-AIR Project in Cork City, Ireland

Marica Cassarino, Roberto Cibin, Dean S. Venables, Denise Cahill, and Kevin Ryan

Background and Introduction

Breathing clean air is considered a fundamental human right (United Nations, 2019, 2022); yet, 99% of the global population is exposed to air quality (AQ) levels that are unsafe for health (WHO, 2021). Air pollution (AP) is a major health threat, increasing the risk of cardiovascular and respiratory diseases, and lung cancer (WHO, 2020), and causing approximately 8 million premature deaths annually worldwide (Lelieveld

M. Cassarino (✉) · R. Cibin (✉)
School of Applied Psychology, University College Cork, Cork City, Ireland
e-mail: mcassarino@ucc.ie

R. Cibin
e-mail: rcibin@ucc.ie

M. Cassarino · D. S. Venables
Sustainability Institute, University College Cork, Cork City, Ireland

D. S. Venables
School of Chemistry, University College Cork, Cork City, Ireland

T. Skillington and A. Setti (eds.), *Climate Change Resilience Across Societal Contexts*, Energy, Climate and the Environment,
https://doi.org/10.1007/978-3-031-95020-9_4

et al., 2023). While Ulrich Beck once claimed that "smog is democratic" (Beck, 1992, p. 36), implying it affects everyone equally, recent studies (e.g. Health Effects Institute, 2024) reveal a disproportionate impact of AP on disadvantaged groups such as minorities, indigenous people, children, older people, and those with pre-existing health conditions. In Ireland, where the BEHAV-I-AIR project is based, AP causes around 1600 premature deaths annually (EPA, 2024), and costs society approximately €2 billion (HSE, n.d.).

AP is closely linked with climate change (Fiore et al., 2015; Orru et al., 2017; Pinho-Gomes et al., 2023). Both share common sources, and each can exacerbate the other. Like AP, climate change affects vulnerable populations the most through greater exposure to risks and fewer available resources to cope (Islam & Winkel, 2017). Both AP and climate change are classified as "wicked" problems, characterized by complex, interconnected causes, with a lack of simple solutions (Kreuter et al., 2004; Rittel & Webber, 1973). Wicked problems are perceived differently by various stakeholders, adding to the challenge of addressing the problem (Interaction Design Foundation—IxDF, 2016b). The complexity of these issues makes it difficult for many to grasp their full scope, understand their causes, and take meaningful action. In Ireland, AP is largely caused by personal and household activities such as solid fuel combustion for heating and motor vehicle emissions; these activities generate levels of pollutants such as particulate matter ($PM_{2.5}$) and nitrogen dioxide (NO_2) that exceed WHO guidelines (EPA, 2024). However, public awareness of AP and its health consequences remains low (Gorman et al., 2023; Oltra & Sala, 2018). This lack of awareness calls for enhanced public communication to foster behavioural change, a priority reflected in Ireland's 2023 Clean Air Strategy (Irish Government, 2023).

D. Cahill
Cork Healthy Cities, Cork City, Ireland

K. Ryan
Air Quality Unit, Environment Section, Cork City Council, Cork City, Ireland

Good communication is essential in raising public awareness about environmental issues, and there is a growing consensus that inclusive and collaborative approaches may prove more effective than top-down efforts (Few et al., 2007; Riley et al., 2021). An inclusive approach helps incorporate diverse perspectives, promotes knowledge sharing, and builds resilience to these challenges. Resilience is defined as the capacity of individuals and communities to adapt to and recover from adverse conditions (APA, n.d.). However, critics argue that mainstream definitions of resilience often place too much responsibility on individuals, ignoring the systemic social, political, and economic factors that shape these challenges (Forsyth, 2023). To address this limitation, there is a shift in focus from individual to collective resilience through collaborative processes such as learning networks (Balint, 2011), researcher-community partnerships (Zandt et al., 2020), and social learning initiatives (Van Epp & Garside, 2016).

The BEHAV-I-AIR (Behaviour and Impacts on Air Quality) project, conducted in Cork City, Ireland, embodies such a collaborative approach. This interdisciplinary, community-engaged initiative aimed to foster AQ-related resilience through engaged public communication. Building on Science and Technology Studies (STS), we conceptualise resilience also as a socio-technical process involving human (e.g. citizens, scientists, policymakers) and non-human (e.g. regulations, technologies, pollutants) actors working together to address AP and climate change. This approach acknowledges the complex dynamics between people, technology, and policies, and focuses on building synergies and fostering trust across these actors towards common goals (Forsyth, 2023; Law, 2023). By adopting a collective, inter-relational understanding of resilience applied to AQ behaviours and practices, BEHAV-I-AIR utilises communication as a tool to co-creating adaptation and mitigation responses within the local context, considering values, power dynamics, and relationships.

In the following sections, we will first introduce the BEHAV-I-AIR project and outline its theoretical framework and main activities. We will then discuss how the project fostered collaboration and trust-building across sectors to engage the public on AP-related issues. Lastly, we will reflect on the key lessons learned, highlighting the challenges and opportunities for building resilient communities dedicated to improving AQ.

BEHAV-I-AIR: AQ Communication to Foster Behavioural and Social Change

Context of the Project

The BEHAV-I-AIR project combines innovations in AQ monitoring with community-engaged design thinking to improve public communication of AP risks and solutions. Launched in Cork City in June 2023, the project is funded by Taighde Éireann Research Ireland (formerly Science Foundation Ireland) under the Healthy "Environment for All" National Challenge Fund. It involves interdisciplinary collaboration between academics in applied psychology, atmospheric chemistry, social sciences, data science, and human–computer interaction, along with local authority representatives and practitioners.

Cork City offers an ideal context for this project. Through its AQ Unit, Cork City Council (the governing local authority) launched Ireland's first local AQ Strategy in 2021. This preceded the national strategy on AQ which was launched in 2023. Cork City Council's AQ strategy outlines the actions that the Council is undertaking to reduce the concentrations of air pollutants in the city area; thereby positively impacting on the quality of life of residents and visitors. The staff within the Council's AQ Unit have collaborated closely with the Centre for Research in Atmospheric Chemistry at University College Cork to advance AP monitoring across the city. One of the deliverables from this collaboration is Ireland's first district scale, real time, particulate matter monitoring network across Cork City. These joint efforts for AQ promotion have also benefited from Cork's designation as a WHO Healthy City (WHO, n.d.), which has enabled over the past 12 years a local growing network of inter-agency collaboration for promoting social and environmental enablers for health. Furthermore, Cork City has established a local network of community and not-for-profit organizations through the local Public Participation Network (PPN) which fosters democratic and inclusive decision-making processes in the local area. This policy and practice context provided a strong foundation for our project team to work on a complex environmental issue through stakeholder engagement and community participation.

In the next sections, we discuss the project's theoretical framework and outline its methodology and associated activities.

An Integrated Theoretical Framework

BEHAV-I-AIR draws on environmental psychology, particularly models of environmental risk perception (Böhm & Tanner, 2013) and Bamberg's (2013) stage model of self-regulated behavioural change. These frameworks helped guide the search for effective strategies to communicate scientific information for behavioural impact. A systematic review of AQ communication strategies (Cibin et al., 2023) helped us to expand our approach and adopt an innovative theoretical framework integrating behavioural science, communication, and socio-ecological models.

Firstly, we consider AQ-related behaviours in light of the Capability, Motivation and Opportunity for Behaviour (COM-B) model (Michie et al., 2011, 2014), which highlights that behaviour change depends on individual capability, motivation, and external opportunities. While COM-B is effective in identifying behavioural intervention pathways, it has been critiqued for focusing primarily on individual factors and neglecting systemic influences. To address this issue, we incorporated a socio-ecological perspective (Nguyen-Trung et al., 2023), which situates individual behavioural change within a broader context of policy, infrastructure, and systemic opportunities (and challenges) that influence action.

Communicating AQ and its health impacts involves complex scientific concepts that must be understood by diverse audiences with varying levels of knowledge, motivation, and capacity to act. To better grasp these interactions, we integrated a Science Communication model, first proposed by Trench (Trench, 2008) and later remodelled by Bucchi and Trench (Bucchi & Trench, 2021a, 2021b). This model considers communication across three different levels of interaction (and thus social and power dynamics) between experts and the public, ranging from top-down expert-to-public dissemination to a two-way dialogue-based consultation, to full public participation in knowledge production and communication.

By combining the COM-B, socio-ecological, and science communication models, we approached AQ communication not just as information dissemination, but as a dynamic interaction among diverse actors, both human and non-human. This perspective allowed our team to explore how different forms of engagement can influence individual, collective, and systemic capabilities, motivations, and opportunities.

Design Thinking, Community Engagement, and Participation

Guided by an integrated theoretical framework, we approached AQ communication as a social innovation aimed at tackling a wicked problem—AP—that creates social and health inequities (Manzini, 2015). To achieve this innovation through stakeholder engagement, and building on our collaborative ecosystem and integrated theoretical foundation, the project development was inspired by Design Thinking. This process allows teams to tackle ill-defined issues while understanding and responding to user needs (Interaction Design Foundation—IDxF, 2016a, 2016b), and also enabling a dynamic interplay between research and design (Stappers & Giaccardi, 2014).

The Design Thinking process typically consists of five phases: Empathise, Define, Ideate, Prototype, and Test. Figure 4.1 provides an overview of these phases within BEHAV-I-AIR. Following the development of our theory of change, we focused on the Empathise and Define phases by conducting stakeholder and process mapping. This effort helped us identify key stakeholders for assessing AQ communication needs. We engaged multiple individuals from diverse sectors, including community, healthcare, policy, business, and academia, through interviews and surveys. Additionally, we organised a public roundtable to

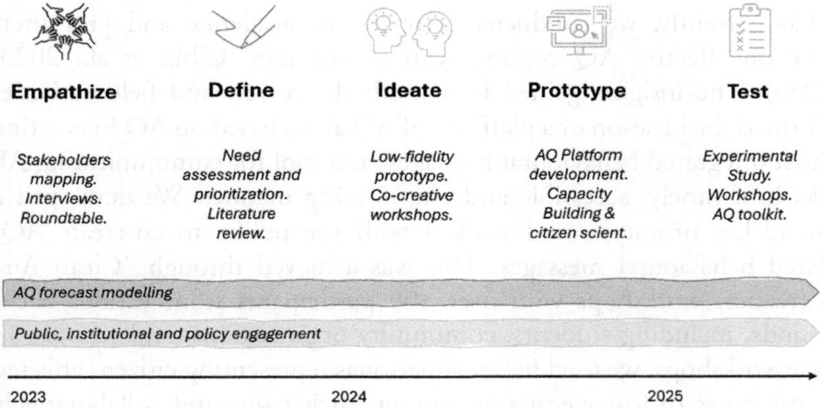

Fig. 4.1 Design thinking process within BEHAV-I-AIR

create a shared vision for clean air (Cassarino & Cibin, 2023) and to mobilise public engagement with our project.

In our consultations, we identified a complex array of 28 communication and structural needs related to promoting public awareness of AQ and empowering less-polluting behaviours (BEHAV-I-AIR project, 2023; Cibin et al., 2025b). Stakeholders observed that existing AQ information was limited, difficult to understand, and perceived as inactionable:

> I went to a local meeting. There was absolutely no awareness at all, and when I talked about the air pollution, people just look blankly. It is really not understood, you know, air pollution is something that happens in Peru or China or somewhere. (Public Health expert)

Many respondents also highlighted infrastructural and policy-related barriers that hinder behaviour change, such as inefficient public transport and inadequate resources for home renovations:

> Some people cannot afford gas and oil, so they use wood. It is cheaper and it works as well and so I think an issue in Ireland is saying "This is a problem" but not giving alternatives. (Young person with asthma)

Concurrently, we conducted a review of academic and grey literature on effective AQ communication strategies (Cibin et al., 2023, 2025a). The insights gained from both the review and field activities informed the ideation of a platform of AQ alerts based on AQ forecasting which integrated behavioural messages, as a tool for communicating AP risks in a timely, accessible and empowering manner. We developed a low-fidelity prototype and worked with the public to co-create AQ-related behavioural messages. This was achieved through "Clean Air" co-creation workshops with over 100 participants from various backgrounds, including students, community organisations, and schools. In these workshops, we used fictional personas representing citizens affected by AP issues to foster empathy among participants and collaboratively develop behavioural messages for the platform. The co-created messages and feedback on the prototype informed a new version of the AQ platform, which, at the time of writing, is undergoing an expert evaluation before being tested to assess its impact on AQ awareness and behaviours.

While we began this project with a socio-technical innovation focus, ongoing engagement with the public and relevant institutions have enabled the project to take unexpected directions and realise the real focus for our work, which is to build synergies and mobilise both the public, organisations and institutions around the issues of AP. This has resulted in a much more diversified communication approach across different media and channels. Our Clean Air workshops also evolved into opportunities for storytelling, using AQ data to foster a sense of agency among participants, ultimately becoming a form of light-touch community interventions. To involve as many people as possible, we adapted the workshops paying specific attention to using communication methods and languages familiar and understandable to the participants. The activities usually took place in the participants' local areas, e.g. community centres, schools and even local community gardens. These efforts have enabled the creation of relationships of trust, which have been crucial to further progress our project activities.

Building Synergies and Fostering Trust for AP Resilience

Wicked problems such as AP and the climate crisis impact society across public health, the environment, social justice and the economy. Our team's focus on collective resilience emerged from an understanding that, while individual behaviours play a part in these problems, solving these complex issues cannot happen by working in silos. As discussed by Chin et al. (2023), achieving positive impacts requires collective efforts at multiple levels to foster a sense of collective ability to adapt and respond to pressures and crises. According to these authors, social trust is a prerequisite of the social processes needed to foster resilience.

In BEHAV-I-AIR, we have striven to achieve trust through intense stakeholder involvement and participatory methodologies. However, building collective resilience to counter AP requires not only the involvement of various stakeholders affected by the issue but also an institutional scaffolding that facilitates action across different system levels. This included particularly our societal partner from Cork City Council, along with key partners in Cork Healthy Cities and the Cork Public Participation Network, and our advisory team of academic experts. The synergies created among these entities anchored our research within local policy and practice, while helping the development of a collaborative ecosystem across multiple societal and organisational levels, as shown in Fig. 4.2. Such ecosystem has been instrumental to develop and diversify the project activities and remit, and to foster concerted efforts to tackle the multifaceted issue of AQ. Importantly, project partners and champions have been crucial in establishing relationships with both local and national stakeholders that are impacted by AP and/or have a stake in promoting positive change.

Engaging with schools, identified by stakeholders as key actors for clean air action, has been a significant achievement for our team. School outdoor spaces often suffer from reduced AQ due to car traffic and idling during drop-off and pick-up times. At the same time, schools provide an ideal educational context for fostering a culture of clean air actions among teachers and students, which can extend to their families. To leverage this opportunity, we conducted Clean Air workshops in

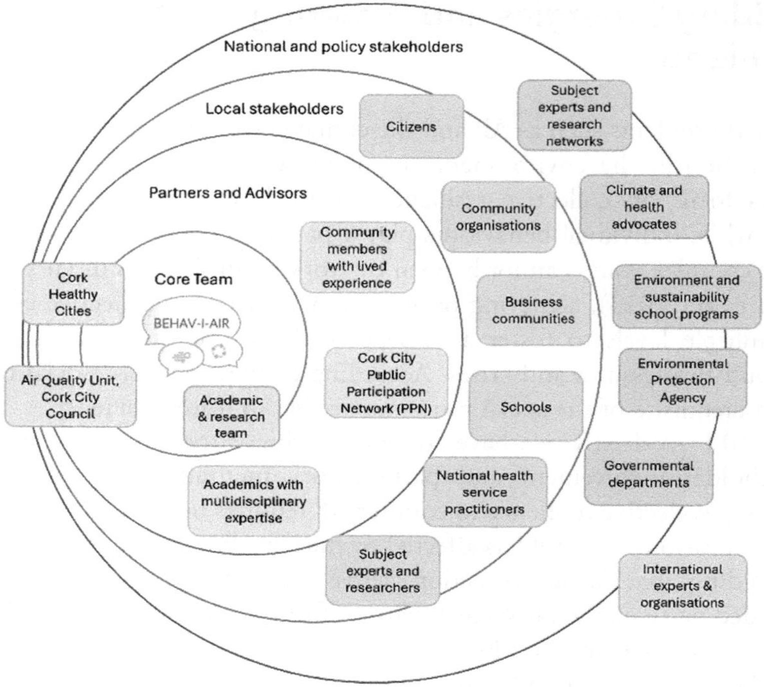

Fig. 4.2 BEHAV-I-AIR collaboration ecosystem and partnerships

both primary and post-primary schools, and collaborated with a team of Schools Health Promotion practitioners in Cork City to design a parent guide to the health risks of engine idling outside schools; this is the practice of leaving vehicle engines running while stationary, which contributes to unnecessary pollution and negative health impacts, particularly for children. The guide was distributed to over 500 schools in the southern area of Ireland, and has become an open resource for members of the public and school communities.

Our team has also engaged in various outreach, advocacy and policy-direct activities, including presentations with health organisations, and participation in public events. We supported the launch of Ireland's first Clean Air Night in Cork with a social media campaign and an opinion piece on solid fuel burning risks. We also worked with the City

Council and Cork Healthy Cities to co-host the Cork Clean Air Day 2024, leading a community event where local champions shared their stories of acting for the promotion of clean air for all. Policy-directed actions include submissions to public consultations and the formation of a working group with the Irish Climate and Health Alliance to draft a position paper on AQ and health. We have also developed relationships with the Air Quality Policy Unit at the Department of Environment, Climate and Communication, the Environmental Protection Agency, national environmental programmes for schools, and health advocacy organisations.

As discussed by Centola (2021), these activities have fostered the creation of "wide bridges", connecting trusted actors, groups and organisations, which can bring about positive change by fostering a sense of shared agency and resilience. While building cross-sectoral and multi-stakeholder synergies and relationships of trust has been, in our case, a stepping stone to promoting mobilisation around AP issues, our experience has shown that this capacity building process is neither easy nor error-proof. A few key learnings have emerged, which we discuss in the following sections together with recommendations for future research and practice in this area.

Negotiating Constraints and Pressures When Working with Communities

Research on environmental and health issues increasingly shows the value of participatory tools and citizen involvement in knowledge production and decision-making. Our work, founded on participatory design thinking, supports the importance of communication strategies that encourage dialogue and participation, fostering a deeper understanding of complex environmental and climate issues. However, this approach is time and resource intensive, often challenging to align with the demands of publicly funded projects. As Centola (2021) explains, social innovations requiring behavioural changes are a form of "complex contagion": unlike simple contagions like diseases, which spread with one contact, complex contagions require multiple interactions and sources of exposure for change to occur. This reflects our own experience of engaging stakeholders with the invisible issue of AP.

Building community synergies has required months of active presence, multiple diversified (at times failed) engagement attempts, and adaptability to the needs and interests of different stakeholders. Demonstrating the value of our work and gaining both public and institutional support has been key. Sustaining these connections is particularly difficult in the current context, where communities and institutions are dealing with multiple societal and environmental crises. This has taught us to be humble in our expectations and appreciative of the time and energy that communities can contribute. While these efforts are essential for sowing seeds of change, their impact is not always measurable through traditional academic metrics, such as publications or immediate societal outcomes. The full impact may not become clear until after the project concludes.

Furthermore, while enhancing public awareness of AP (or climate) issues through engaging communication is necessary to promote behavioural and social changes, there is no single "silver bullet" communication strategy that fits all needs or that guarantees behavioural impact. Our stakeholder consultations revealed different AP communication preferences and needs based on individual and contextual circumstances. We have tried to address in our project through different activities and collaborations. Yet, due to time and funding constraints we have been faced with important (and at times difficult) choices about which strategies to prioritise, such as focusing communication efforts on schools.

Navigating these constraints has been facilitated by our interdisciplinary team and the support of our project partners. Working within this ecosystem encouraged reflective practice, which has been essential for creativity in our work. For instance, during our fieldwork, we observed a strong need for tools and resources that communities could use to enhance their understanding of AP independently. In response, a new development in our project is the creation of a capacity-building toolkit for AQ-related action as an open resource for communities and organisations to identify their own communication-based solutions for the promotion less polluting behaviours.

While these participatory approaches to environmental communication present challenges, they also offer meaningful opportunities for long-term change by fostering deeper engagement and understanding within communities.

Build It and They'll Come? A Critical Look on Who Participates (and Who Doesn't)

Ensuring and diversifying stakeholder participation is a key ingredient for the success of community-engaged research, but also possibly the most difficult challenge faced by initiatives like BEHAV-I-AIR working on complex environmental and health issues.

We have been fortunate to reach different groups and stakeholders across the local community thanks to our partners and connectors. However, many people who attended project events already had a certain sensitivity to AP issues, and we had less representation from groups who are socially vulnerable or those who are mis- or dis-informed about the topic. Both these groups were identified in our initial stakeholder mapping as those who may benefit the most from learning about the risks and solutions associated with AP. We also observed that levels of awareness and understanding of AP was very limited even among highly educated individuals and groups; this may impede meaningful action.

These observations highlight the challenges of developing inclusive and effective communication strategies for the promotion of sustainable behaviours. Firstly, participatory efforts may be not sufficient to reach the disconnected element of society, who usually are not part of the public debate. Secondly, issues with environmental and digital literacy can act as barriers to meaningful participation in social innovations -let alone acceptance of the same. Thirdly, working on communication of environmental issues requires a careful consideration of dynamics related to misinformation, hate speech, trolling and digital injustice.

Importantly, individual behavioural changes related to AP (or other climate issues), particularly for socially vulnerable groups, are conditioned by constraints and opportunities in their social and structural contexts (Evans, 2007). Indeed, in our consultations, many stakeholders pointed out policy or infrastructural barriers to assist the movement away from polluting behaviours, and to a certain level of disillusion towards "the system" and sense of disenfranchisement. Adaptive and resilient behaviours can only occur in social, political and economic systems that are both resilient and adaptable themselves. These systemic barriers may nullify any well-intentioned efforts to foster behavioural change.

Some Recommendations

The learnings from the BEHAV-I-AIR project highlight some challenges in trying to foster a sense of collective resilience for climate-related issues such as AP. Even though the solutions to problems outlined above are by no means simple, our own experience suggests some potential ways forward to address them in future research and practice:

- The creation of networks with civil society organisations that act as intermediaries in reaching people usually not involved in the public debate and that can foster buy-in and participation.
- Using environmental communication not only as a strategy targeted to individual behaviours but also as a means for public and institutional mobilisation through framing environmental problems such as AP as also health, social and climate justice issues.
- Combining digital and analogic communication channels to those experiencing digital divides to access and benefit from information.
- Promoting positively framed and action-oriented messages around AP (or other climate-related) issues which highlight the need to change behaviours while also providing solutions, many of which can be a suite of simple steps for positive change. The work we have done with schools, our outreach activities and the upcoming capacity building toolkit are the first steps in this direction.

These learnings are highly transferable to environmental and climate-related issues beyond AP and may prove helpful for future work aimed at supporting climate resilience in a feasible and sustainable way. The BEHAV-I-AIR project provides an example of an innovative approach to making AQ communication inclusive and empowering, informed by an integrated theoretical framework and participatory design thinking. The project has benefited from a focus on capacity building and collaboration at multiple societal levels to foster social trust and collective resilience, as opposed to looking at individual behaviours only. This approach is not without its challenges. However, it can bring meaningful benefits to both researchers, policymakers, practitioners and the community at large.

References

APA. (n.d.). *APA dictionary of psychology—Resilience*. Retrieved October 21, 2024, from https://dictionary.apa.org/resilience

Balint, P. J. (2011). *Wicked environmental problems: Managing uncertainty and conflict*. Island Press.

Bamberg, S. (2013). Applying the stage model of self-regulated behavioral change in a car use reduction intervention. *Journal of Environmental Psychology, 33*, 68–75. https://doi.org/10.1016/j.jenvp.2012.10.001

Beck, U. (1992). *Risk society: Towards a new modernity* (1st edition). SAGE Publications Ltd.

BEHAV-I-AIR project. (2023, October 26). *BEHAV-I-AIR needs statements*. https://behaviair.org/wp-content/uploads/2024/03/behav-i-air_needsstatements_v31.10.23.pdf

Böhm, G., & Tanner, C. (2013). Environmental risk perception. In L. Steg & J. I. M. de Groot (Eds.), *Environmental psychology: An introduction* (pp. 13–25). John Wiley & Sons Ltd.

Bucchi, M., & Trench, B. (2021a). Rethinking science communication as the social conversation around science. *Journal of Science Communication, 20*(3), Y01. https://doi.org/10.22323/2.20030401

Bucchi, M., & Trench, B. (Eds.). (2021b). *Routledge handbook of public communication of science and technology* (3rd ed.). Routledge. https://doi.org/10.4324/9781003039242

Cassarino, M., & Cibin, R. (2023). *Clean air day roundtable—Report of key results and actions*. Open Science Framework. https://osf.io/, https://osf.io/h6xu3

Centola, D. (2021). *Change: How to make big things happen*. Little, Brown Spark.

Chin, I., Joerin, J., & Schubert, R. (2023). *Strengthening social resilience—The Importance of Trust* (FRS Working Paper, #7). https://ethz.ch/content/dam/ethz/special-interest/dual/frs-dam/documents/FRS_Working_Paper_7_Strengthening%20Social%20Resilience.pdf

Cibin, R., Ciolfi, L., Dockray, S., Venables, D. S., & Cassarino, M. (2023). *BEHAV-I-AIR–Protocol for a Rapid Review on the role of communication interventions on air quality behavioural change*. Open Science Framework. https://osf.io/, https://osf.io/4gczp

Cibin, R., Horgan, L., Ciolfi, L., Dockray, S., Murphy, G., Venables, D. S., & Cassarino, M. (2025a). Promoting air quality-related awareness and behavioural change: A rapid review of communication interventions integrating behavioural, science communication and socio-ecological perspectives. *Environmental Challenges*. https://doi.org/10.1016/j.envc.2025.101237

Cibin, R., Venables, D. S., & Cassarino, M. (2025b). Promoting urban air quality resilience through communication: A stakeholder needs mapping and prioritisation study informed by design thinking. *City and Environment Interactions*. https://doi.org/10.1016/j.cacint.2025.100209

EPA. (2024). *Air quality in Ireland 2023*. Environmental Protection Agency. https://www.epa.ie/publications/monitoring--assessment/air/air-quality-in-ireland-2023.php

Evans, K. (2007). Concepts of bounded agency in education, work, and the personal lives of young adults. *International Journal of Psychology, 42*(2), 85–93, Article 2.

Few, R., Brown, K., & Tompkins, E. L. (2007). Public participation and climate change adaptation: Avoiding the illusion of inclusion. *Climate Policy, 7*(1), 46–59. https://doi.org/10.1080/14693062.2007.9685637

Fiore, A. M., Naik, V., & Leibensperger, E. M. (2015). Air quality and climate connections. *Journal of the Air & Waste Management Association, 65*(6), 645–685. https://doi.org/10.1080/10962247.2015.1040526

Forsyth, T. (2023). *Climate change adaptation and resilience: sociotechnical and knowledge dimensions*. Routledge.

Gorman, G., Hellebust, S., Venables, D., Ryan, K., & Cassarino, M. (2023). Introducing forecast-based public health warnings to promote engagement with air quality risk: A survey of citizens' attitudes in Cork Ireland. *Journal of Risk Research, 26*(6), 594–609. https://doi.org/10.1080/13669877.2023.2170453

Health Effects Institute. (2024). *State of global air 2024. Special Report*. Health Effects Institute. https://www.stateofglobalair.org/resources/report/state-global-air-report-2024

HSE, H. P. S. C. (n.d.). *Health effects of air pollution.* Retrieved August 9, 2024, from https://www.hpsc.ie/a-z/environmentandhealth/airquality/healtheffectsofairpollution/

Interaction Design Foundation—IDxF. (2016a, May 25). *What is design thinking?* Interaction Design Foundation—IDxF. https://www.interaction-design.org/literature/topics/design-thinking

Interaction Design Foundation—IxDF. (2016b, June 4). *What are wicked problems?* Interaction Design Foundation—IDxF. https://www.interaction-design.org/literature/topics/wicked-problems

Irish Government, (gov.ie). (2023, April 26). *Clean air strategy*. https://www.gov.ie/en/publication/927e0-clean-air-strategy/

Islam, N., & Winkel, J. (2017). *Climate change and social inequality* (ST/ESA/2017/DWP/152; DESA Working Paper No. 152). OECD iLibrary. https://doi.org/10.18356/2c62335d-en

Kreuter, M. W., De Rosa, C., Howze, E. H., & Baldwin, G. T. (2004). Understanding wicked problems: A key to advancing environmental health promotion. *Health Education & Behavior, 31*(4), 441–454. https://doi.org/10.1177/1090198104265597

Law, T. (2023). *Climate change adaptation and resilience—Introduction.* Routledge.

Lelieveld, J., Haines, A., Burnett, R., Tonne, C., Klingmüller, K., Münzel, T., & Pozzer, A. (2023). Air pollution deaths attributable to fossil fuels: Observational and modelling study. *BMJ, 383*, Article e077784. https://doi.org/10.1136/bmj-2023-077784

Manzini, E. (2015). *Design an introduction to design for social innovation.* . The MIT Press. https://doi.org/10.7551/mitpress/9873.001.0001

Michie, S., Atkins, D. L., & West, P. R. (2014). *The behaviour change wheel: A guide to designing interventions.* Silverback Publishing.

Michie, S., van Stralen, M. M., & West, R. (2011). The behaviour change wheel: A new method for characterising and designing behaviour change interventions. *Implementation Science, 6*(1), 42. https://doi.org/10.1186/1748-5908-6-42

Nguyen-Trung, K., Saeri, A. K., Zhao, K., Boulet, M., & Kaufman, S. (2023). *A brief introduction to a socio-ecological COM-B (SeCOM-B): A behaviour change framework response to wicked problems.* https://doi.org/10.31219/osf.io/4x6wa

Oltra, C., & Sala, R. (2018). Perception of risk from air pollution and reported behaviors: A cross-sectional survey study in four cities. *Journal of Risk Research, 21*(7), 869–884. https://doi.org/10.1080/13669877.2016.1264446

Orru, H., Ebi, K. L., & Forsberg, B. (2017). The interplay of climate change and air pollution on health. *Current Environmental Health Reports, 4*(4), 504–513. https://doi.org/10.1007/s40572-017-0168-6

Pinho-Gomes, A.-C., Roaf, E., Fuller, G., Fowler, D., Lewis, A., ApSimon, H., Noakes, C., Johnstone, P., & Holgate, S. (2023). Air pollution and climate change. *The Lancet Planetary Health, 7*(9), e727–e728. https://doi.org/10.1016/S2542-5196(23)00189-4

Riley, R., de Preux, L., Capella, P., Mejia, C., Kajikawa, Y., & de Nazelle, A. (2021). How do we effectively communicate air pollution to change public attitudes and behaviours? A review. *Sustainability Science, 16*(6), 2027–2047. https://doi.org/10.1007/s11625-021-01038-2

Rittel, H. W. J., & Webber, M. M. (1973). Dilemmas in a general theory of planning. *Policy Sciences, 4*(2), 155–169. https://doi.org/10.1007/BF01405730

Stappers, P., & Giaccardi, E. (2014, January 1). *Research through design.* Interaction Design Foundation - IxDF. https://www.interaction-design.org/literature/book/the-encyclopedia-of-human-computer-interaction-2nd-ed/research-through-design

Trench, B. (2008). Towards an analytical framework of science communication models. In D. Cheng, M. Claessens, T. Gascoigne, J. Metcalfe, B. Schiele, & S. Shi (Eds.), *Communicating Science in Social Contexts: New models, new practices* (pp. 119–135). Springer Netherlands. https://doi.org/10.1007/978-1-4020-8598-7_7

United Nations. (2019). *Clean air and human rights* (A/HRC/40/55; Special Rapporteur on Human Rights and the Environment, David Boyd). https://www.ohchr.org/en/special-procedures/sr-environment/clean-air-and-human-rights

United Nations. (2022, July 26). *The human right to a clean, healthy and sustainable environment*. UN. https://digitallibrary.un.org/record/3982508

Van Epp, M., & Garside, B. (2016). *Solving 'wicked' problems: Can social learning catalyse adaptive responses to climate change?* International Institute for Environment and Development. https://www.jstor.org/stable/resrep02746

WHO. (n.d.). *WHO European healthy cities network*. Retrieved October 21, 2024, from https://www.who.int/europe/groups/who-european-healthy-cities-network

WHO. (2020). *World health statistics 2020: Monitoring health for the SDGs, sustainable development goals*. https://www.who.int/publications/i/item/9789240005105

WHO, W. H. O. (2021). *WHO global air quality guidelines: Particulate matter (PM2.5 and PM10), ozone, nitrogen dioxide, sulfur dioxide and carbon monoxide*. https://www.who.int/publications-detail-redirect/9789240034228

Zandt, S. V., Masterson, J. H., Newman, G. D., & Meyer, M. A. (2020). *Engaged research for community resilience to climate change*. Elsevier.

5

Resilient Cities of the Future: What Will They Look Like?

Tadhg E. MacIntyre, Raegan Thompson, Niamh Ní Chongahile, and Elaine Gallagher

Introduction

Cities are our future. In Europe currently there are over 800 cities (urban conurbations with a population of greater than 50,000 inhabitants) accounting for almost 76% of the population with global figures currently at 56%. That percentage is expected to reach nearly 82% in the

R. Thompson
Illinois School of Professional Psychology, National Louis University, Chicago, IL, USA
e-mail: raeganb55@icloud.com

T. E. MacIntyre (✉) · N. Ní Chongahile · E. Gallagher
Innovation Value Institute, Maynooth University, Maynooth, Ireland
e-mail: Tadhg.Macintyre@mu.ie

N. Ní Chongahile
e-mail: nichonghaileniamh@gmail.com

E. Gallagher
e-mail: elaine.gallagher@mu.ie

T. Skillington and A. Setti (eds.), *Climate Change Resilience Across Societal Contexts*, Energy, Climate and the Environment,
https://doi.org/10.1007/978-3-031-95020-9_5

EU and to 70% globally by 2050 (UN, 2011). Urbanisation has been characterised by many challenges and threats. Collectively, these issues have been referred to as posing a 'wicked problem.' This term refers to problems which are deemed intractable, as they have fuzzy boundaries, and are characterised as messy and devious systems of interacting problems (Termeer et al., 2019). Labelling a problem 'wicked' doesn't mean it cannot be solved but can create an understanding that multiple solutions may be required and efforts to solve one aspect may thus create other problem. It is the nexus of challenges in terms of climate change, urban pollution, and land use change that can have a detrimental impact on human and environmental health. In this chapter, we pose a counterpoint to this pessimistic viewpoint and explore how cities can be integral to preserving and promoting our right to a clean, healthy, and sustainable environment (UN, 48/13, 2021). To support our case, we will review the problems typically associated with urban living, the context for a new paradigm and the opportunities that it brings us. Subsequently, the role biological principles including resilience have in elucidating our understanding will be outlined. Next, two case studies and a series of recommendations will be followed by a brief conclusion.

'Wicked Problems'

Three major challenges, climate change, urban health and biodiversity loss are not entirely unrelated, as we will discuss. For example, in many contexts increased temperatures may lead to the 'Urban Heat Island' effect which exacerbates the impact of air pollution (Nieuwenhuijsen, 2024) despite success across the EU in reducing the levels of ambient air pollution. Threat from increased temperatures poses a major risk for airborne particles by causing air turbulence that increases particulate matter (PM2.5, PM10) in micro-climates (Lungman et al., 2024). Approximately one-third of heat-related mortality is attributable to human-induced climate change with high temperatures claiming over 60,000 lives in Europe alone in 2022 (Nieuwenhuijsen, 2024). One key metric to understand the impact of sub-optimal temperatures is how many days are expected above 35 degrees Celsius in major cities. Models looking at 2051–2011 predict that 72% of European cities (411 out of 571), are projected to experience an increase in a maximum temperature increase during heatwaves of at least 10 °C (Guerreiro et al., 2018). These temperatures will result in diminished opportunities for outdoor exercise, active travel and recreation for many vulnerable groups (e.g. older adults and young children) particularly where relative humidity is also high. The co-hazards of air and noise pollution both have an impact upon human health and well-being, with higher levels of air pollution associated with increased symptoms of depression, anxiety, and stress (Leontjevaite et al., 2024). Air pollution also undermines soil health and biodiversity, while leaf size index and other measures of environmental quality can serve as bio-indicators of climate stress and air pollution (European Environment Agency, 2022). Nature in the city, as we will see can be part of the solution to ensure cities are more resilient.

The New Urban Agenda

Despite the aforementioned threats there is an increase on their standing that urban nature and nature-based solutions have a role not only in mitigation and adaptation but in promoting resilient cities (MacIntyre et al., 2019). This thinking has emerged in recent decades from what can only be described as a paradigm shift led by a wave of initiatives, strategies, and legislative tools. The UN New Urban Agenda, UN Decade of Restoration, New European Bauhaus, EU Green Deal and EU Biodiversity Strategy 2030, augmented by the recent Nature Restoration Regulation all provide a firm foundation for the proliferation of new solutions that showcase cities as potential beacon's of sustainable and healthy living. This bright and optimistic vision of cities of the future, one conducive to the preservation of essential ecosystems and natural habitats, can ensure citizens, communities and diverse groups thrive with nature. The key question we need to address is how do we get there?

Multidimensional Health Approaches

In the past ten years new models have emerged to help us understand the interconnections between human and environmental health. The planetary health perspective and one health approaches offer unified frameworks which can aid are understanding of the interlinkages when applying a multidimental health perspective. In this chapter we will focus on the planetary health approach because it is offers a more comprehensive approach to understand humans interactions in the urban context. One health, by comparison, offers a broader scope in evaluating risks of zoonosis (transmission of diseasees across species), for example. The planetary health approach aims to understand how environmental changes can change natural surroundings and the human reaction to these changes from a global to a local perspective (Pongsiri et al., 2019).

Biological Principles

Planetary health and mulitidimensional health models pose interlinkages between human and environmental health across a range of boundaries. If this is the case one should look to biological principles to shed light on our understanding of living in a dynamic and changing urban ecosystem. Three such principles, variation, homeostatis, metamorphosis and resilience are consistent in how they mediate our behaviour across contexts. Variation is an argument utilised to explore individual differences, while operating under a general framework that can categorise populations of human functions or behaviour (Kosslyn et al., 2002). For too long psychologists have overlooked variation in biological processes such as processing information, hormones, our immune system, and the structure of neurons. Once research operates under the impression that humans may have unique biological processes there may be more detailed descriptions about the human psyche.

Metamorphosis, or our own unique developmental trajectories, can help us appreciate how in nature acorns become trees, a copse of trees becomes a forest and a forest can support a vast ecosystem in terms of biodiversity. Nature in this form, a topic we will return to shortly, also provides an area for recreation, social interactions and exercise for nearby residents. Resilience, a construct we will now explore, has been understood as both a human attribute and a property of natural systems, whereby it traditionally refers to the speed of return to the equilibrium state of an ecosystem.

What is Resilience?

The origin of the word resilience is from the Latin word 'resilire' meaning to spring back. However, this isn't the full understanding from a psychological perspective. The term resilience has been plagued with conceptual inconsistencies, particularly when considered across different contexts (Bryan et al., 2019). A comprehensive review defined resilience as 'encompassing the capacity to maintain regular functioning through diverse challenges or to rebound through the use of facilitative resources (Bryan et al., 2019). This conceptualization is aligned with the conservation of resources theory (Hobfoll, 2011) whereby resilience is conceived of as a state rather than a trait. Resilience can thus be viewed as a form of flexible self-regulation enhancing our ability to bounce forward instead of simply bouncing back encouraging psychological growth.

Applying Resilience

In practice, this means that resilience may be more clearly understood as a state-like phenomenon involving processes of adaptation and growth from stressors. Fluctuations in specific psychological resources may influence resilience capacities to deal with stressors and evidence from organisational psychology suggests that we can activate resources through social support, proactive coping and a range of adaptive strategies (Butler et al., 2021). The events varying from minor short-term (e.g. daily hassles), to life-stress events and long-term adversities, which can resonate with the impacts of climate change for example. The American Psychological Association, in 2008 summarised the psychological impacts of climate change include eco-anxiety (chronic or severe worry about the environment and future of the planet), ecological grief, and climate-related trauma and stress. This can include direct impacts (e.g. chronic disaster syndrome and PTSD), indirect impacts (e.g. vicarious trauma) and more enduring stress (e.g. slow onset trauma). Resilience is not merely an antidote but a necessity given these potential impacts and the global impact of climate change and other threats which often asymetrically impact cities. However, cities and re-natured cities can be a key to unlock the solution.

Resilient Nature?

Given the synergies between natural systems and the human construct of resilience the potential to cultivate resilience through nature-based experiences had until recently been overlooked (Ingulli & Lindbloom, 2013). Recently, Smith et al. (2019) elaborated on the role of nature-based activities to foster resilience from a developmental perspective. Evidence on the role of nature-based interventions more generally suggests that have a moderate but consistent positive impact on self-reported well-being in the workplace, for example (Gritzka et al., 2020). Given the aforementioned global and EU led initiatives which support cities through re-naturing and nature-based solutions, and community based initiatives for well-being and nature engagement, there is now hope that this new lens for citizens can support more healthy and sustainable lives and a pathway towards achieving the UN SDG's for 2030. Amongst the fundamental requirements are that nature in a city is sufficient in scale and accessibility to support citizens use of nature for health and well-being. The WHO recommendation for urban greenspace is that citizens should have access to 0.5 hectares within 5 minutes of their residence. A seminal paper in this field, estimated that if this recommendation was met in Europe up to 43,000 lives would be saved annually (Barboza et al., 2021). Nature is thus a heath determinant for urban citizens, and two brief case studies now convey how nature interactions can support resilience.

Case Studies

1. **The Woodlands for Health Nature-Based Intervention:** This programme is a nature-based green exercise intervention for client users who have a lived experience of a mental health issue. It is coordinated by multiple key actors in the outdoor nature sector, Coillte, the Irish Forestry Agency, with Sport Ireland agency 'Get Ireland Walking and NGO Mental Health Ireland.' It involves guided walks through our forests landscapes over 12 weeks and has proliferated across Ireland. Preliminary evaluation showed positive impacts in terms of well-being, nature connection and social cohesion (Mental Health Ireland, 2020).
2. **GoGreen Routes Technology Supported Nature-Based Resilience Training:** This mobile Application was part of an evidence-based 6-week intervention which aimed to enhance resilience, wellbeing and environmental stewardship in the GoGreen Routes EU Horizon Project. The nature-based dimensions were designed to strengthen the user's connection for nature and the outdoors. The programme itself was housed within a tailored mobile application and embedded tasks addressing a specific resilience-related theme every week. Each theme contained a written informational component, as well as a selection of carefully crafted nature-based activities designed to promote psychological resilience and bring the user closer to nature. Pilot study findings reported in the project showed a statistically significant improvement in psychological resilience and positive feedback from the users based on their experience of the application.

Recommendations

Researchers can benefit from applying a multidisciplinary approach to psychological resilience and seek to understand how resilience interventions can be embedded in natural settings. Practitioners can use nature-based interventions, perhaps applying nature not just as a backdrop to the therapeutic context (i.e. outdoor therapy) but put nature at the heart of the intervention, and explore ecological techniques (Delaney, 2020). At the community level, it is vital that resilience as a concept is not imposed upon citizens, notably amongst diverse groups who may diverge on their locus of control when it comes to interpreting stressful and traumatic climate-related events. A bottom-up approach, whereby citizens co-create their own definition of resilience can augment the academic understanding and promote a shared mental model to activate citizens and ensure a participatory approach to community resilience.

Conclusions

There is potential Resilience, not the property of any one individual or setting, is dependent upon our own interconnections with others and nature alike. Re-thinking resilience can be part of the solution to ensuring that cities with their economies of scale and efficiencies support a more resilient future with benefits for both human and environmental health and well-being.

Funding Acknowledgement This research has been supported by the European Union under the Horizon Europe research framework under grant agreement no. 101137209. The sole responsibility for the content of this website lies with the GoGreen Next project and in no way reflects the views of the European Union.

References

Barboza, E. P., Cirach, M., Khomenko, S., Iungman, T., Mueller, N., Barrera-Gómez, J., Rojas-Rueda, D., Kondo, M., & Nieuwenhuijsen, M. (2021). Green space and mortality in European cities: A health impact assessment study. *Lancet Planet Health, 5*(10), e718–e730. https://doi.org/10.1016/S2542-5196(21)00229-1. PMID: 34627476.

Bryan, C., O'Shea, D., & MacIntyre, T. (2019). Stressing the relevance of resilience: A systematic review of resilience across the domains of sport and work. *International Review of Sport and Exercise Psychology, 12*(1), 70–111. https://doi.org/10.1080/1750984X.2017.1381140

Butler, C. G., O'Shea, D., & Truxillo, D. M. (2021). Adaptive and proactive coping in the process of developing resilience. In P. D. Harms, P. L., Perrewé & S.-H.(D) Chang (Ed.), *Examining and exploring the shifting nature of occupational stress and well-being* (*Research in Occupational Stress and Well Being, Vol. 19*) (pp. 19–46). Emerald Publishing Limited. https://doi.org/10.1108/S1479-355520210000019002

Delaney, M. (2020). *Nature is Nurture: Counseling and the Natural World.* Oxford University Press.

European Environment Agency. (2022, November 24). *Impacts of air pollution on ecosystems—European Environment Agency.* www.eea.europa.eu. https://www.eea.europa.eu/publications/air-quality-in-europe-2022/impacts-of-air-pollution-on-ecosystems

Guerreiro, S. B., Dawson, R. J., Kilsby, C., Lewis, E., & Ford, A. (2018). Future heat-waves, droughts and floods in 571 European cities. *Environmental Research Letters, 13*(3), Article 034009. https://doi.org/10.1088/1748-9326/aaaad3

Gritzka, S., MacIntyre, T. E., Dörfel, D., Baker-Blanc, J. L., & Calogiuri, G. (2020). The effects of workplace nature-based interventions on the mental health and well-being of employees: A systematic review. *Front. Psychiatry, 11*, 323. https://doi.org/10.3389/fpsyt.2020.00323

Hobfoll, S. E. (2011). Conservation of resources theory: Its implication for stress, health, and resilience. In S. Folkman (Eds.). *The Oxford handbook of stress, health, and coping,* (pp. 127–147). Oxford University Press.

Ingulli, K., & Lindbloom, G. (2013). Connection to nature and psychological resilience. *Ecopsychology, 5*(1), 52–55. https://doi.org/10.1089/eco.2012.0042

Kosslyn, S. M., Cacioppo, J. T., Davidson, R. J., Hugdahl, K., Lovallo, W. R., Spiegel, D., & Rose, R. M. (2002). Bridging psychology and biology: The analysis of individuals in groups. *American Psychologist, 57*(5), 341–351. https://doi.org/10.1037/0003-066X.57.5.341

Leontjevaite, K., Donnelly, A., & MacIntyre, T. E. (2024). Air pollution effects on mental health relationships: Scoping review on historically used methodologies to analyze adult populations. *Air, 2*(3), 258–291. https://doi.org/10.3390/air2030016

Lungman, T., et al. (2024). The impact of urban configuration types on urban heat islands, air pollution, CO_2 emissions, and mortality in Europe: A data science approach. *The Lancet Planetary Health, 8*(7), e489–e505.

MacIntyre, T. et al. (2019). Nature-based solutions and interventions in cities. A look ahead. In A.A. Donnelly & T. E. MacIntyre (Eds.). *Physical activity in natural settings* (pp. 335–338). Routledge. https://https://doi.org/10.4324/9781315180144-18

Mental Health Ireland. (2020). Woodlands for health evaluation report: Nature-based interventions and well-being a preliminary study. Mental Health Ireland. https://www.mentalhealthireland.ie/wpcontent/uploads/2025/05/Woodlands-for-Health-Evaluation-Report-2020_FINAL.pdf

Nieuwenhuijsen, M. J. (2024). Climate crisis, cities, and health. *The Lancet, 404*(10463), 1693–1700. https://doi.org/10.1016/S0140-6736(24)01934-2

Pongsiri, M. J., Bickersteth, S., Colón, C., DeFries, R., Dhaliwal, M., Georgeson, L., Haines, A., Linou, N., Murray, V., Naeem, S., Small, R., & Ungvari, J. (2019). Planetary health: From concept to decisive action. *The Lancet Planetary Health, 3*(10), e402–e404. https://doi.org/10.1016/s2542-5196(19)30190-1

Smith, S., Hunt, E., O'Brien, M., MacIntyre, D. (2019). Growing resilience through connecting with nature. In A. A. Donnelly & T. E. MacIntyre (Eds.). *Physical Activity in Natural Settings*: Green and Blue Exercise. Routledge.

Termeer, C. J. A. M., Dewulf, A., & Biesbroek, R. (2019). A critical assessment of the wicked problem concept: Relevance and usefulness for policy science and practice. *Policy and Society, 38*(2), 167–179. https://doi.org/10.1080/14494035.2019.1617971

UN (2021). General assembly resolution A/RES/73/284: Decade of restoration to prevent, halt and reverse the degradation of ecosystems worldwide. United Nations.

UN HABITAT (2011). *Cities and climate change: Global report on human settlements*. Routledge.

Part II

Resilience Building Across Generations

6

A Dynamic Approach to Climate Resilience in Ageing

Annalisa Setti, Andrea Stitzel, Giovanni Ottoboni, and Niamh McCarthy

Ageing and Climate Change: Definitions

The WHO states that '*resilience is related to processes and skills that result in good individual and community health outcomes, in spite of negative events, serious threats and hazards [...]. More recently this definition has been broadened to include the desire and urgency to strengthen the resilience*

A. Setti (✉)
School of Applied Psychology and Environmental Research Institute, University College Cork, Cork, Ireland
e-mail: a.setti@ucc.ie

A. Stitzel
School of Health Science and Social Work, Carinthia University of Applied Sciences, Villach, Austria

G. Ottoboni
Department of Psychology, University of Bologna, Bologna, Italy

N. McCarthy
School of Psychology, University College Cork, Cork, Ireland

T. Skillington and A. Setti (eds.), *Climate Change Resilience Across Societal Contexts*, Energy, Climate and the Environment,
https://doi.org/10.1007/978-3-031-95020-9_6

of social systems, including health [...]' (World Health Organization. Regional Office for 2017, p. 3). Key to this concept is the identification of three levels at which resilience can be measured and promoted, the individual level, the community level and the systems level. The need for the system in which the individual is included to be proactive in promoting resilience is key to the concept of resilience-transforming capacity, which highlights the dynamic nature of resilience. Resilience, in line with the Sustainable Development Goals, requires ***absorptive, anticipatory,*** *and* ***adaptive*** *capacities and well as the ability for individuals, communities and systems to transform* (United Nations, 2020). These aspects of resilience are applicable to the individual, who can prepare for future negative circumstances, to be less impacted by them and react better to the impact but also transform to diminish the likelihood of a major adverse impact and of more minor, but more often occurring events. For example, an older person may prepare for a mega storm about to hit their city, but they can also adapt to the constantly changing climate with adaptive changes, e.g. structural (to their home), in terms of lifestyle and psychological transformation, in order to improve resilience.

To adapt to the unavoidable impacts of climate change and to become climate resilient by 2050, the EU Mission Adaptation (European Commission, 2021a) sets ambitious emission reduction targets, but simultaneously acknowledges that emission reduction will '*have little effect on the overall trajectory of global warming … and that substantial adaptation efforts would still be required*' (European Commission, 2021a, p. 3). The proposed long-term vision is to develop climate-resilient European societies through a systemic approach that targets all levels and sectors. Increasing local resilience responses is central to the strategy, as this level reveals the most evident unequal exposure of individuals to the effects of climate change, leading to further aggravation of already existing vulnerabilities. The EC (European Commission, 2021a, p. 9) points out that '*the impacts of climate change are not neutral. Men and women, older people, persons with disabilities, displaced persons, or socially marginalised have different adaptive capabilities. Adaptation measures need to consider their situation.*' Therefore, the envisioned 'just resilience' strategy has a strong social component aiming at engaging and empowering European citizens to become active agents in climate adaptation.

A further cross cutting priority in the EU Strategy on Adaptation to Climate Change (2021) are nature-based solutions benefitting adaptation, mitigation, disaster risk reduction, biodiversity, as well as individual health (co-benefits).

In response to the increasing aging population, fifteen years ago, the European Commission embraced policy on *healthy and active ageing* which aims to empower older adults to remain in control of their lives for as long as possible. The goal is to continue to live independently and contribute meaningfully to their society and the economy (Europe 2020, A strategy for smart, sustainable and inclusive growth [European Commission, 2010, 2021b]). However, this policy will continue to face barriers caused by climate change. Extreme weather events, rising temperatures, and natural disasters are expected to negatively impact older adults in their physical and psychological vulnerability. Still, there is a significant gap in understanding how older adults psychologically adapt to climate change. As highlighted in two recent the Lancet articles (Mavrodaris et al., 2021; Mavrodaris et al., 2022), the pursuit of the UN Sustainable Development Goals goes hand in hand with reducing age-related inequalities in a changing environment. While older people may be a group at risk of social inequalities due to climate change, here we reject the victimizing view of older people, which could be potentially associated with ageism, as was often the case during the Covid-19 pandemic.

In this chapter, we will explore resilience to climate change, in particular its mental health and wellbeing effects, taking a dynamic approach to ageing, considering ageing as part of our lifespan development. We propose that resilience building occurs at micro and macro level and overlaps with other aspects of ageing, such as promoting a fulfilling life, minimizing the negative impacts of cognitive and physical changes, promoting social interaction and a sense of meaning. We adopt the view of older people as active agents in their interaction with other societal agents, and we stress the importance to include older people in the climate mitigation as well as in the climate adaptation discourse.

Climate Change Effects on Wellbeing and Mental Health

An in-depth narrative review on the current climate change effects on mental health and emotional wellbeing suggested several negative impacts on individuals (Lawrance et al., 2022). Two broad distinctions are made: 'direct' exposure to acute or chronic climate-related events and 'indirect' observations of long-term climate change impacts. Direct effects on mental health are primarily related to air pollution (increasing the risk for depression, ADHD, and schizophrenia), extreme heat (linked with aggressive behavior, suicide, and sleep disturbances), and natural disasters (post-traumatic stress disorder, general anxieties, depression, and increased suicide rates). Indirect effects, stemming from an awareness of impending changes—also called 'eco-anxiety/eco-distress, climate or ecological grief, or solastalgia'—are linked to symptoms such as 'feelings of loss and grief, despondency, suffering despair, helplessness, hopelessness and [...] sleep disturbances' (White et al., 2023, p. 8). Lawrance et al. (2022) describe a continuum from direct to indirect experiences, as acute shocks can, over time, lead to increased indirect psychological distress.

Older people, along with children, are particularly vulnerable to both the direct and indirect mental health effects of climate change. Regarding direct effects, both extreme heat and temperature fluctuations are linked to mental ill-health among older adults, with vulnerability being especially high among those with pre-existing mental health conditions or low socio-economic status (White et al., 2023; Gebhardt et al. 2023). Although evidence regarding the indirect effects of climate change on older persons seems to be missing from the literature, it can be deduced that ecological grief may be prominent among older persons, as 'ecological grief has been found to affect those with strong ties to a particular place' (Lawrance et al., 2022, p. 14).

While research on resilience has developed in recent years, the specific psychological resilience of older adults in relation to climate change remains largely unexplored. Currently, there is no comprehensive review that combines the collective findings of psychological resilience in older adults regarding climate change. This gap is particularly concerning given the increasing frequency and severity of climate-related disasters, which threaten the physical safety, mental health and emotional well-being of the older population. Some insights can be drawn from available studies addressing resilience in ageing. The literature focuses mainly on coping with adversity provoked by climate change. Some effective strategies to cope with climate change, either chronic effects or acute stressors have been identified, such as focus on meaningful relationships with people (Adams et al., 2014), with nature (Dennis & Stock, 2024; Khalil et al., 2025), pets (Thompson et al., 2024) and pro-environmental action (Khalil et al., 2025). Self-care and developing emotion regulation mechanisms through psycho-education (Elzohairy et al., 2024) have also proven effective. However more work is needed to inform policy and practice.

Now imagine a couple, they live independently, and they are in their mid-sixties, they own their house. Both retired, the couple enjoy relatively good health and have some social ties in the town where they live. When their third grandchild was born, they decided to remodel their home to transfer their bedroom and bathroom to the ground floor. The decision is mainly related to considerations on their potential mobility difficulties in the years to come, as well as to create more space for grandchildren visiting, however some climate change-related considerations come into play. They worry about heat waves becoming more persistent over the years; the ground floor is cooler than the top floor and air conditioning can be easily installed too. The remodeling affords the opportunity to prevent inconvenience, increase health risks prevention and the sense of agency and preparedness towards potential issues arising.

A different situation is experienced by a lady living in an apartment on the ground floor, which she owns, in a medium-sized city. She is in her mid-seventies, with good social ties, although she is widowed, and one of her two children has passed away recently due to illness. She is a member of the local knitting club, and she used to meet her friends there, although less so after the loss of her son. She is also less confident in going out, as she noticed her memory fails at times. She is worried about the changing climate, particularly the strength of the storms as of late, but does not feel that there is an urgency to take any steps in her own daily life.

Finally, think about a gentleman living in a state-owned accommodation, he has no family and retired early due to issues with his leg. In his neighborhood, new families have moved in, that he has no particular contact with. He used to go out with colleagues but nowadays his main contact is with a befriending phone service provided by a charity. He also receives cooked meals from the City Council every second day. He finds great enjoyment in visiting the local park and he is very saddened by the loss of plants and animals that he sees in this place dear to him; he is concerned about rumors he heard about plans to transform the park introducing walkways with pavements and lighting.

These three examples portray individuals living in an urban context in the Global North. These individuals are personas (Pruitt & Grudin, 2003) in that they represent certain individual and contextual circumstances, and their situation is captured in a specific moment in time. They display different levels of functioning, connection and autonomy, therefore different levels of 'successful' (Rowe & Kahn, 1997) and 'active' ageing in the traditional definitions (European Commission, 2021b). Their wellbeing is also differently affected by climate change, in relation to immediate and more temporally distant threats. This is within the context of a different perception of their own quality of life and wellbeing. One could argue that the couple in the first scenario has taken the opportunity to increase their ability to respond to acute (heat waves) and chronic threat as well as future threat (increasing heat waves and potentially diminished physical function), therefore displaying both the absorptive and anticipatory dimensions of resilience. Their socio-economic status and availability of services facilitate them in displaying such resilience. The second scenario portrays a lady who has diminished wellbeing due to a tragic life event. Although she has adapted to the event and goes on with her life, her social ties have weakened, and she could be potentially extremely vulnerable to climate-related events further increasing her isolation, such as it has been during the Covid-19 pandemic. Her socio-economic status gives her some protection against climate-related events, e.g., owning an apartment where she feels comfortable. This persona displayed resilience to her life events, and she is absorbing her changing cognitive status, however she is not proactively seeking to change her current lifestyle or prepare for a change in circumstances, internal or environmental. The last persona is heavily reliant on social support provided by agencies, public or private, his social interactions have decreased after retirement; he enjoys the local nature available, which is free and accessible, however he worries about gentrification and suffers from solastalgia, witnessing the degradation of the local natural habitat. The social services contribute to his adaptive resilience to retirement, as well as the local physical environment, which, however, has a dual role, given the experience of biodiversity loss. One could imagine that, given the opportunity, he may consider joining a local Biodiversity group.

In the three examples, all the individuals display resilience to life circumstances that have shaped their current wellbeing and quality of life, their physical and social immediate environment present opportunities and obstacles, and, in the last example, we know that there are systems in place to support the individual (meals delivery and friendly calls), however we do not know how resilient the systems themselves are to climate-related threats. This is an important point to highlight. While in this chapter we will focus on individuals in their ageing process, as highlighted by Lawrance et al. (2022), the psychological pathways in which climate change affects individual wellbeing are embedded in multiple systems and their interactions (Dahlgren & Whitehead, 2021).

Ageing Well in a Time of Climate Change: Ageing as a Multifaceted Phenomenon

When exploring climate resilience in the older population, we consider two aspects: how the ageing process itself is conceptualized (Waddell et al., 2024), as the very definition of ageing impacts the approach to assessing and supporting resilience; and how impacts, particularly the negative ones, of climate change on wellbeing and quality of life are conceptualized and potentially considered to inform interventions (Lawrance et al., 2022).

In the current ageing literature, it is emphasized that older people are not a homogeneous group, and wide differences exist between individuals of the same demographic age, which is reflected in the response to different demands, starting from their neurophysiology (Cabeza, 2002). Cognitive ageing (Ram et al., 2011) and frailty (Fried et al., 2001; Rockwood & Mitnitski, 2007) potentially affect resilience; for example, the lady in the second scenario may withdraw from going out of her house partly because she is afraid of getting lost or of being seen as forgetful by others. The specific environmental context may co-determine individual differences; the same older person may be different, in terms of mental health, behaviors and outcomes, depending on the situation in which they operate. Ageing stereotypes, for example, impact how people react to different situations, and internalized stereotypes and negative views of ageing affect proactivity towards health-related behaviors (Fawsitt & Setti, 2017), as well as a range of performances (cognition, vision), and overall quality of life (Fawsitt et al., 2022). Ageing stereotypes (or stereotypes of other kind such as low socio-economic status), for example, may contribute to preventing the older man in the latter scenario to actively engaging with a local Biodiversity group, which could increase his social network and increase the sense of agency over the changing environment. The first couple shows environmental mastery. Two life events, retirement and a grandchild being born, have fostered a positive adaptation, which is potentially increasing their resilience and current wellbeing.

Ageing well is characterized by a series of dimensions, depending on the theoretical framework, summarized by Waddell et al. (2024) which include adaptation, resilience, purpose and meaning in life, self-acceptance, environmental mastery, personal growth and expertise, speaking to the appraisal of the individual of their own situation as an ageing person, as well as to the absence or level of disease and function. These dimensions should be taken into consideration when working with older individuals to support them in increasing their climate resilience.

Climate change does not necessarily represent a-priori a higher risk for the older person than for a younger one, as there are multiple intra and inter-individual differences with ageing. To understand the maladaptive outcomes of climate change for the older person, it is necessary to take into consideration their expectations, their meaning of life, their own perception of any limitations they may experience, and the social context in which the person is inserted. People responses in the specific situations, and their capacity to maintain a good level of well-being, will be the resultant of multifactorial agents and factors. Already Baltes (1996) proposed that the ways in which people respond to life changes (including ageing and disease) is rooted in their bio-psycho-social history, in the context in which they live, in the presence and characteristics of the services that are available where they live, and in the political choices of those who govern the socio-economic-health development of the region and the nation, and in the ways in which all these actors interact (Ottoboni et al., 2021). According to Baltes and Baltes (1990), Selective Optimization with Compensation (SOC) refers to three elements. Selection occurs when there is a choice to be made, and potentially a narrowing of possibilities in ageing. In the examples above, the older lady who does not feel confident to go out of her apartment choses to stay home. Optimization, the second component of the SOC model, focuses on doing the best with available resources to achieve goals and adapt to situations. It emphasizes maximizing compensatory strategies. Baltes and Baltes (1990) link optimization to the richness of one's environment, a poor context may hinder selecting options or implementing external strategies, forcing reliance on internal solutions. Optimizations through internal strategies may be promoted or hindered by the opportunities that person could avail of throughout their life, as under-stimulation may have also undermined the possibility of creative solutions. Compensation, the final component of the SOC model, addresses strategies used to counterbalance declining abilities. It involves adopting tools or methods, such as glasses for reading or moving to the ground floor in your house, like the couple in the first example, to maintain goals despite losses. However, compensation can be constrained by factors like financial or logistical challenges, which may require new solutions. For example, if the couple could not afford to modify their home, they could have reduced the impact of heating with air conditioning. Compensation includes both material alternatives, like physical tools, and immaterial strategies, such as psychological coping mechanisms. For example, the man in the third example, builds psychological resilience by cultivating his connectedness with nature.

The Role of Life Changes and Transitions

As ageing is a process in which changes, both individual and contextual, occur, therefore ageing well in times of climate change requires considering the temporal dimension of ageing, which is often overlooked. Dennis et al. (2023) adopted a knowledge brokering approach with researchers and practitioners to highlight some key principles to promote wellbeing in the ageing population through green infrastructure, which, we propose, are applicable to resilience building interventions more broadly. Any intervention should be person-centered and empower the individual, it should consider their values, and it needs to consider the temporal dimension, both in terms of the individual and the timeframe of the effects one aims to obtain. Dennis et al. (2023) point to the importance of life transitions in generating opportunities for the individual to change. In the scenarios above, the birth of a grandchild, loss of a child and retirement are major life transitions which were associated with coping strategies having an impact, direct or indirect, on climate resilience. This idea is not new in studying individual change. The Habit Discontinuity Hypothesis (Verplanken & Roy, 2016) utilized in behavioral change theory, proposes to determine opportunities for small and big changes capitalizing on life changes that organically break individual habits. For example, moving house could break the habit of driving and incentivize the individual to walk to the shops. More broadly, borrowing from the field of counseling psychology with the Life Transitions model (Schlossberg, 1981), we propose that life changes, big or small, should be considered in determining whether they support or hinder individuals' resilience. These significant life transitions are nested into broader slower changes, such as cognitive decline or age-related sensory loss, and distinct from them, however, impact people's life; for example dealing with the negative psychological consequences of having to stop cycling, or the positive consequences of adopting a new hobby.

Lifestyle, Lived Environment, and Resilience

Many of the activities helping the individual to age well are also potentially resilience-building, directly or indirectly. For example, being physically active can support cognitive health, which, in turn, is a building block of emotion regulation and coping. Participating in coordinated pro-environmental action has potentially a positive effect on socialization and on climate mitigation (Lawrance et al., 2022), as well as combating ageing stereotypes. The STAC model (Scaffolding Theory of Aging and Cognition), in its revised form (STAC-R) (Reuter-Lorenz & Park, 2014), examines how cognitive abilities, environmental inputs, and psychophysiological reserve interact to shape individual responses to challenges, including climate-related demands. It emphasizes the brain's adaptive capacity, both structurally and functionally, using 'scaffolding' to support individuals in coping with life events. Neural resources are influenced by enriching factors (e.g., education, physical activity) or depleting ones (e.g., stress, depression, low socioeconomic status). Compensatory mechanisms, like neuronal plasticity and strengthened connectivity, counter aging effects. The STAC-R emphasizes a lifelong dynamic interaction between enriching and depleting influences on neural resources. It advocates for proactive interventions (e.g., education, physical activity) to bolster cognitive resilience, offset age-related challenges and build resilience for future challenges. In the previous examples, the possibility for the couple to play with their grandchildren, or the older man walk in the park, constitute activities that support body and brain's health. However, even resilience building activities like those could be associated with negative wellbeing effects of climate change. For example, the couple could develop climate anxiety for the future of their grandchildren and the older man could develop climate grief and solastalgia for the biodiversity loss he witnesses. Therefore, it is important to consider these possibilities and address them with positive messaging about climate action and interventions, where appropriate (Lawrance et al., 2022).

Another theoretical framework that is at the background of this chapter is the idea of person-environment fit. First introduced to the area of Gerontology by Lawton and Nahemow (1973) as a way to express behavior as a function of the interaction between personal characteristics and environmental stressors, we consider it here a more broad conceptualization aiming at understanding fit as function of the compatibility between individual and environmental characteristics (Kaplan, 1983), which can be physical, social and cultural (see Zhang et al., 2025, for a critical review). People live in neighborhoods, cities, regions; they have relationships with family, friends, and acquaintances, they also live in a socio-cultural environment (Bronfenbrenner, 1979). While the person-environment fit idea has been revised multiple times, it still lacks a clear temporal dimension (Zhang et al., 2025). Zhang et al. (2025) suggest that a life cycle approach to person-environment fit could be more productive as *fit* is likely to change over time. For example, if the park did not exist in the vicinity of the older man's house, he would not have available an environment that fits his love for nature and fosters his health by availability of natural spaces to walk in. Availability of biodiverse natural environments could be particularly resilience building for those who are neurologically more environmentally sensitive (Greven et al., 2019). Environmental Sensitivity is an umbrella term encompassing differences in noticing and responding to the environment, which presents both disadvantages (low resilience) and advantages depending on the situation. Connecting with nature (Dunne et al., 2024) and contact with nature (Mariani Wigley et al., 2025) have a particularly positive impact on these individuals and are associated with higher flourishing in middle aged and older highly sensitive people (Carroll et al., 2025).

Take-Aways

Interventions aimed at fostering (climate) resilience in older people should consider their individual history, their goals and the particular time in their life in which the intervention occurs. Climate change should be seen in this context as an amplifier and climate resilience not as an entirely new concept. Nevertheless, the use of the term climate resilience seems appropriate, as it brings with it a heightened sense of urgency and raises the question of how older people, in particular, can contribute to finding answers to the needed societal transformation.

Older people are not a homogenous group, while dealing with the effects of ageing in the context of the impeding climate crisis, they maintain the right to autonomy and agency over the way they aim to age and live. Therefore, it is important to **move away from a dichotomy between those who are or are not climate resilient, to understand when, where and how different ageing individuals display resilience** and how to foster that.

This dynamic approach to ageing in times of climate change promotes **policy implementation,** especially on the local level that, while acknowledging the particular risks that older adults are exposed to due to climate change, can cater for such diversity of experiences and goals. Local communities should actively involve older people when developing climate resilience strategies and interventions to match the needs of this growing population group. Older people's knowledge and perspective is paramount to creating conducive environments fostering their own and their community's resilience. Specific consideration should be given to the question of how **co-designing** with vulnerable older persons is possible. Such participatory actions also have the added benefit—when well-designed—to strengthen social ties, increasing a sense of belonging and personal value, hereby fostering resilience.

A prerequisite to implement **co-creative processes**, however, is **capacity building** among older people themselves, those who work directly with them but also other stakeholders in the community. Often ageing stereotypes are in the way of seeing older people as active agents in the face of a changing climate. Fostering spaces for older people to be **visible** in their **environmental action will co-benefit** the individual as well as foster societal transformation. Activities can span from small individual to collective intergenerational actions, with volunteering opportunities to be offered in such a way that they are not only physically but also emotionally and culturally barrier-free for older people. Capacity building also must extend to the question of how to buffer mental ill-health of older people in the face of acute climate-related shocks and increase emotional wellbeing vis-a-vis long-term climate stressors. Psycho-educational programs specifically for/with older people and those working with them are a promising way forward. Raised awareness can be expected to lead to greater communication and acceptance of the mental health effect of climate change, especially among those older adults most vulnerable (due to multiple vulnerabilities such as pre-existing mental health conditions, chronic diseases, lower socio-economic status, missing social networks).

Nature connectedness, i.e., **connection with non-human organisms** (plants, gardens, pets, wild animals)**, constitutes a potential network of relationships** that should not be underestimated when considering climate resilience in ageing. Connectedness with nature varies due to neurological differences, personal history, access to nature, and current needs. We propose that the accessibility of a biodiverse natural environment can foster climate resilience and cultivate meaning in life and climate action in ageing.

References

Adams, J., Pascal, J., & Dickson-Swift, V. (2014). Spirituality and aging in place: The impact of extreme climatic conditions on domestic gardening practice. *The International Journal of Aging and Human Development, 80*(1), 10–26. https://doi.org/10.1177/0091415015591107

Baltes, M. M. (1996). *The many faces of dependency in old age. The many faces of dependency in old age.* Cambridge University Press.

Baltes, Paul B., & Margret M. Baltes. (1990). Psychological perspectives on successful aging: The model of selective optimization with compensation. In *Successful aging: Perspectives from the behavioral sciences* (pp. 1–34). Cambridge University Press.

Bronfenbrenner, U. (1979). *The ecology of human development*. Harvard University Press.

Cabeza, R. (2002). Hemispheric asymmetry reduction in older adults: The HAROLD model. *Psychology and Aging, 17*(1), 85–100. https://doi.org/10.1037/0882-7974.17.1.85

Carroll, S., O'Brien, A., Lionetti, F., O'Reilly, A., & Setti, A. (2025). Flourishing as a highly sensitive person: a mixed method study on the role of nature connectedness and chaotic home environment. *Front. Psychol. Sec. Personality and Social Psychology, 16*. https://doi.org/10.3389/fpsyg.2025.1480669.

Dahlgren, G., & Whitehead, M. (2021). The Dahlgren-Whitehead model of health determinants: 30 years on and still chasing rainbows. *Public Health, 199*, 20–24. https://doi.org/10.1016/j.puhe.2021.08.009. https://www.sciencedirect.com/science/article/pii/S003335062100336X.

Dennis, M. K., & Paul and Stock. (2024). "You're asking me to put into words something that i don't put into words".: Climate grief and older adult environmental activists. *Journal of Gerontological Social Work, 67*(3), 281–296. https://doi.org/10.1080/01634372.2023.2259942

Dennis, M., Barker, A., Anderson, J., & Ashton, J. C., Cavan, G., Cook, P. A., French, D., Gilchrist, A., James, P., Phillipson, C., Tzoulas, K., Philip Wheater, C., Wossink, A., & Lindley, S. (2023). Integrating knowledge on green infrastructure, health and well-being in ageing populations: Principles for research and practice. *Ambio, 52*(1), 107–125. https://doi.org/10.1007/s13280-022-01765-5. https://doi.org/10.1007/s13280-022-01765-5

Dunne, H., Lionetti, F., Pluess, M., & Setti, A. (2024). Individual traits are associated with pro-environmental behaviour: Environmental sensitivity, nature connectedness and consideration for future consequences. *People and Nature, 6*(2), 586–597. https://doi.org/10.1002/pan3.10581

Elzohairy, N. W., Abd El Fatah, N. K., & Abdelwahab Khedr, M. (2024). Effectiveness of psycho-educational program on climate change distress and risk perception among older adults. *Geriatric Nursing, 55*, 35–43. https://doi.org/10.1016/j.gerinurse.2023.10.017

European Commission. (2010). *A strategy for smart, sustainable and inclusive growth.* Communication from the commission EUROPE 2020. 3.3.2010 COM(2010).

European Commission. (2021a, February 24). Forging a climate-resilient Europe—The new EU Strategy on Adaptation to Climate Change. In *Communication from the Commission to the European Parliament, the Council, the European Economic and Social Committee and the Committee of the Regions.*

European Commission. (2021b). The Green Paper on Ageing. In *COM (2021) 50 final.*

Fawsitt, F., Dockray, S., & Setti, A. (2022). Regulatory focus and perceptions of ageing: Exploring the connections. *Aging & Mental Health, 26*(7), 1451–1459. https://doi.org/10.1080/13607863.2021.1952550

Fawsitt, F., & Setti, A. (2017). Extending the stereotype embodiment model: A targeted review. *Translational Issues in Psychological Science, 3*, 357–369. https://doi.org/10.1037/tps0000136

Fried, L. P., Tangen, C. M., Walston, J., Newman, A. B., Hirsch, C., Gottdiener, J., Seeman, T., Tracy, R., Kop, W. J., Burke, G., & McBurnie, M. A. (2001). Frailty in older adults: Evidence for a phenotype. *Journals of Gerontology. Series a, Biological Sciences and Medical Sciences, 56*(3), M146–M156. https://doi.org/10.1093/gerona/56.3.m146

Gebhardt, N., van Bronswijk, K., Bunz, M., Müller, T., Niessen, P., & Nikendei. (2023). Scoping Review zu Klimawandel und psychischer Gesundheit in Deutschland – Direkte und indirekte Auswirkungen, vulnerable Gruppen, Resilienzfaktoren. *Journal of Health Monitoring* (S4), 132–161. https://doi.org/10.25646/11650

Greven, Corina U., Lionetti, F., Booth, C., Aron, E. N., Fox, E., Schendan, H. E., Pluess, M., Bruining, H., Acevedo, B., Bijttebier, P., & Homberg, J. (2019). Sensory Processing Sensitivity in the context of Environmental Sensitivity: A critical review and development of research agenda. *Neuroscience & Biobehavioral Reviews, 98*, 287–305. https://doi.org/10.1016/j.neubiorev.2019.01.009.

Kaplan, S. (1983). A model of person-environment compatibility. *Environment and Behavior, 15*(3), 311–332. https://doi.org/10.1177/0013916583153003

Khalil, M., Mahfouz, I., Shaala, R. S., Sayed Mousa, E. F., Ali Zoromba, M., & Ramadan Atta, M. H. (2025). Examining the associations between emotionally charged reactions toward climate change and self-care, quality of life among older adults, coping mechanisms, and pro-environmental practices. *Geriatric Nursing, 61*, 353–363. https://doi.org/10.1016/j.gerinurse.2024.11.013.

Lawrance, E. L., Thompson, R., Vay, J. N. L., Page, L., & Jennings, N. (2022). The impact of climate change on mental health and emotional wellbeing: a narrative review of current evidence, and its implications. *International Review of Psychiatry, 34*(5), 443–498. https://doi.org/10.1080/09540261.2022.2128725

Lawton, M. P., & Nahemow, L. (1973). Ecology and the aging process. In *The psychology of adult development and aging* (pp. 619–674). American Psychological Association.

Mariani Wigley, I. C., Nazzari, S., Pastore, M., Provenzi, L., & Barello, S. (2025). The contribution of environmental sensitivity and connectedness to nature to mental health: Does nature view count? *Journal of Environmental Psychology, 102*, 102541. https://doi.org/10.1016/j.jenvp.2025.102541.

Mavrodaris, A., Lafortune, L., & Brayne, C. E. (2022). The future longevity: designing a synergistic approach for healthy ageing, sustainability, and equity. *The Lancet Healthy Longevity, 3*(9), e584-e586. https://doi.org/10.1016/S2666-7568(22)00145-3.

Mavrodaris, A., Mattocks, C., & Brayne, C. E. (2021). Healthy ageing for a healthy planet: Do sustainable solutions exist? *The Lancet Healthy Longevity, 2*(1), e10–e11. https://doi.org/10.1016/S2666-7568(20)30067-2

Ottoboni, G., Chirico, I., Povolná, P., Dostálová, V., Holmerová, I., Janssen, N., Dassen, F., de Vugt, M., Sánchez-Gómez, M. C., García-Peñalvo, F., Franco-Martin, M. A., & Chattat, R. (2021). Psychosocial care in dementia in European higher education: Evidence from the SiDECar ("Skills in DEmentia Care") project. *Nurse Education Today, 103,* Article 104977. https://doi.org/10.1016/j.nedt.2021.104977

Pruitt, J., & Grudin, J. (2003). Personas: practice and theory. In *Proceedings of the 2003 conference on Designing for user experiences.* https://doi.org/10.1145/997078.997089.

Ram, N., Gerstorf, D., Lindenberger, U., & Smith, J. (2011). Developmental change and intraindividual variability: Relating cognitive aging to cognitive plasticity, cardiovascular lability, and emotional diversity. *Psychology and Aging, 26*(2), 363–371. https://doi.org/10.1037/a0021500

Reuter-Lorenz, P. A., & Park, D. C. (2014). How does it STAC up? Revisiting the scaffolding theory of aging and cognition. *Neuropsychology Review, 24*(3), 355–370. https://doi.org/10.1007/s11065-014-9270-9

Rockwood, K., & Mitnitski, A. (2007). Frailty in relation to the accumulation of deficits. *Journals of Gerontology. Series a, Biological Sciences and Medical Sciences, 62*(7), 722–727. https://doi.org/10.1093/gerona/62.7.722

Rowe, J. W., & Kahn, R. L. (1997). Successful aging. *The Gerontologist, 37*(4), 433–440. https://doi.org/10.1093/geront/37.4.433

Schlossberg, Nancy K. (1981). A model for analyzing human adaptation to transition. *The Counseling Psychologist, 9*(2), 2–18. https://doi.org/10.1177/001100008100900202. https://journals.sagepub.com/doi/abs/.

Thompson, C., Dilworth, T., James, S., Haynes, K., Batterham, M., & Halcomb, E. (2024). The self-care of older australians during bushfires and COVID-19: A qualitative study. *Disaster Medicine and Public Health Preparedness, 18,* e219. https://doi.org/10.1017/dmp.2024.96. https://www.cambridge.org/core/product/EA7811CB22625D9F59BF9BC1EF23B78B.

United Nations. (2020). *United Nations common guidance on helping build resilient societies.* (UN).

Verplanken, B., & Roy, D. (2016). Empowering interventions to promote sustainable lifestyles: Testing the habit discontinuity hypothesis in a field experiment. *Journal of Environmental Psychology, 45*, 127–134. https://doi.org/10.1016/j.jenvp.2015.11.008. https://www.sciencedirect.com/science/article/pii/S0272494415300487.

Waddell, C., Van Doorn, G., Power, G., & Statham, D. (2024). From successful ageing to ageing well: A narrative review. *The Gerontologist, 65*(1). https://doi.org/10.1093/geront/gnae109.

White, B. P., Breakey, S., Brown, M. J., Smith, J. R., Tarbet, A., Nicholas, P. K., & Ros, A. M. V. (2023). Mental health impacts of climate change among vulnerable populations globally: An integrative review. *Annals of Global Health, 89*(1), 66. https://doi.org/10.5334/aogh.4105

World Health Organization. Regional Office for, Europe. (2017). *Strengthening resilience: A priority shared by health 2020 and the sustainable development goals.* World Health Organization. Regional Office for Europe (Copenhagen). https://iris.who.int/handle/10665/372345.

Zhang, Z., Wenda, Z., Suihan, Z., Yaxiao, C., Xinpeng, W., Yuri, F., & Nobuaki and Furuya. (2025). Person-environment fit theory in built environment: A scoping review. *Journal of Asian Architecture and Building Engineering,* 1–17. https://doi.org/10.1080/13467581.2024.2373824

7

Like I Couldn't Turn into an Eco-Warrior Overnight, You Know? Imaginaries of Climate Change

Angela Veale, James O'Rourke, and Jasmine Byrne

Introduction

On the 9th of August 2021, the United Nations (UN) Intergovernmental Panel on Climate Change (IPCC) published the sixth Assessment Report (AR6) of Working Group I (WGI), titled *'The Physical Science Basis'*. According to this report, "It is unequivocal that human influence has warmed the atmosphere, ocean and land" (IPCC, 2021, p. 2). According to Masson-Delmotte et al. (2021), if global temperatures increase by 2 degrees Celsius higher than pre-industrial temperatures, there will be universal calamity. Tripathy and colleagues (2023) discussed the devastating impacts already being felt around the world as a result of climate change, including prolonged heatwaves and droughts resulting in

A. Veale (✉) · J. O'Rourke
School of Applied Psychology, University College Cork, Cork, Ireland
e-mail: a.veale@ucc.ie

J. Byrne
Independent Researcher, West Haven, CT, United States

T. Skillington and A. Setti (eds.), *Climate Change Resilience Across Societal Contexts*, Energy, Climate and the Environment,
https://doi.org/10.1007/978-3-031-95020-9_7

crops and livestock perishing, poor air quality, and increased frequency and intensity of extreme weather events, such as wildfires and flooding. However, despite widespread dissemination and acceptance of these findings, climate change continues to worsen (IPCC, 2021), and humanity has yet to implement the changes necessary to mitigate and prevent further climate change (Hornsey & Fielding, 2020). Ireland offers an interesting case study in this context. Ireland is vulnerable to climate change due to its location on the Atlantic coast, likely leaving Ireland exposed to rising sea levels, coastal erosion, flooding, and risks to the agriculture and forestry sectors due to increasing temperatures (Flood et al., 2020). Yet, while nearly all (95%) of Irish people are aware of climate change and 89% feel climate change is personally important to them (Environmental Protection Agency, Ireland, 2024), there can be a disconnect between awareness and support for climate action. Although coastal flooding may severely impact coastal cities such as Dublin in coming years, there is no significant public support to construct the necessary flood defences (Laino & Iglesias, 2023).

As the severity of environmental consequences continues to increase, so too does the prevalence of climate change anxiety or eco-anxiety, which the American Psychological Association (APA) and ecoAmerica (2017) define as "a chronic fear of environmental doom. (p. 68)". A systematic study that examined negative emotions about climate change found that worry and climate anxiety were the most prevalent negative emotion; this was even more so the case in studies that engaged exclusively with young people when compared to other adult cohorts (Martin et al., 2023). Hajek and König (2022) found that climate change anxiety is most prevalent in young people under thirty years of age, likely due to young people expecting to live through climate change for a longer time than their older counterparts. Reyes et al. (2023), identified a significant correlation between climate change anxiety and overall mental health difficulties in a "Gen Z" cohort (N = 433), and concluded that climate change is a risk factor for developing mental health difficulties. A Canadian survey of 1000 Canadian young people (aged 16–25) explored climate-related emotions and found that 56% of young people reported they felt afraid, sad, anxious, and powerless, 75% reported that climate change impacts their overall mental health, and 73% reported negative perceptions of the future (Galway & Field, 2023). The findings of this single country study are remarkably consistent with those of a global survey of 10,000 young people aged 16–25 years, in ten countries (1000 per country), across five continents (Hickman et al., 2021). This found that 84% of young people were moderately to very worried about climate change, while nearly half (45%) noted climate-related feelings negatively impacted daily life and functioning. Three quarters (75%) reported that they think the future is frightening.

An incapacity to perceive a positive future may have implications for the capacity to engage in imaginative processes which are often future-oriented. As imagination can be a precursor to, and play a powerful role in, mobilising action (Zittoun & Gillespie, 2016a), it is important to explore young people's imagination with respect to climate change. In this chapter, we report on a series of climate cafés with young people, to explore their imaginings about the climate future.

There is some concern within the literature that talking about climate change may itself have a negative psychological impact. Kricorian and Turner (2022) found that greater exposure to climate change information, more frequent discussions about climate change with friends and family, being younger, and holding a belief that climate change will soon have a personal impact, predicted greater eco-anxiety. Similarly, those enrolled in environment classes were found to have higher stress and anxiety levels compared to non-participants (Kelly, 2024). Yet, the relationship between climate change anxiety and action remains unclear. There are some suggestions that climate change anxiety may mobilise climate actions as individuals may be motivated to engage in pro-environmental behaviours as a positive adaptive coping response (Innocenti et al., 2023; Kricorian & Turner, 2022; Martin et al., 2022; Ojala & Bengtsson, 2019). Alternatively, such anxiety may have a detrimental effect on taking action as a sense of hopelessness and helplessness may result in eco-paralysis (Heeren & Asmundson, 2023; Innocenti et al., 2023; Stanley et al., 2021). Clayton (2020) advocates for distinguishing between adaptive and maladaptive climate anxiety. She argues it is important not to pathologise climate anxiety, as it may be a rational response to the available evidence base and its future orientation may motivate individuals to positive climate action.

Imagination is an important psychological mediator between emotion and action. In a rich imaginative landscape, individuals draw on culturally available material, drawn from their social and semiotic worlds (Vygotsky, 2004). This gives rise to creative activity oriented towards the future, thus giving rise to the possibility of altering behaviour in the present to bring about such a desired future. Imaginative practices facilitate scenario building (Yusoff & Gabrys, 2011), which links orientations to actions. In a study of Iranian children (aged 9–13) exploring imaginings of the future of the Anthropocene, utopian environmental images were associated with children who considered themselves to have efficacy in changing the future of the environment, while dystopian images came from "children who were inactive and who did not accept any responsibility" (Demneh & Darani, 2020, p. 269). Muglan (2022) argues, as a species, we are suffering from an imaginary crisis, a deep unease affecting our capacity to imagine, whether social or political, and argues for a need to revitalise our collective imagination. In this chapter, we explore Irish college students imaginings of climate change, the cultural and material resources drawn on in imagination and their orientation to the future.

Conceptually, this paper draws on Zittoun and Gillespie's model of imagination (Zittoun & Gillespie, 2016b; 2018). In this model, imagination is triggered by some internal or external event: for example, boredom, or as in the case of the study reported here, an invitation to imagine the climate future. This trigger results in a rupture from the current "real" moment and activates imaginary processes in which experientially, one "loops into" an imaginative loop, in which culturally available semiotic and material resources are drawn on and arranged in innovative or creative ways (Zittoun & Gillespie, 2016b; Zittoun & Cerchia, 2013). Key dimensions of imagination in this model include sphere of experience (proximal or distal to daily contexts and realities), orientation (the imaginative loop can be oriented towards the past or future), and plausibility, whereby the imaginative loop may range along a continuum from the plausible to the implausible, that is to have a "more or less degree of likelihood or possible realisation" (Zittoun & Gillespie, 2018, p. 18), before eventually coming back to present experience.

Cultural resources are the symbolic and material resources that make up the contents of a person's imagination, create meaning in the world, and allowing them to communicate with others (Zittoun & Gillespie, 2016b). In the case of this study, what are the symbolic and material (physical) resources that form the contents of a student's imaginings about climate change? And what symbolic and material resources might they draw on as they imagine solutions to the climate crisis? By examining these cultural resources in imagination, it is hoped that an understanding can be created as to how social and cultural factors shape participant perceptions of climate change and how they imagine solutions to it.

To transition towards sustainability, collective action driven by shared imagination is essential (Zittoun & Gillespie, 2018). This will require imagining new ways of being in everyday life, that replace current societal norms that are destroying our planet. Only then, can individuals and collectives begin to transform their communities by bringing those imagined ways of being into reality.. In this way, imagination, in particular collective imagination, can be a powerful psychological resource for creating societal transformation, by establishing possible actions and values for the individual as well as the collective (Zittoun, 2017).

Methodology

To explore the ways in which young people imagine climate change, the study invited young people to participate in a 'Climate Café'. A Climate Café is "an open, inclusive space for people to get together to talk and act on climate change. Climate Cafés are informal where everyone is welcome to join the conversation and get involved. Cafés can bring people together from across communities, workplaces, and campuses to focus on solutions" (Climate Café, n.d.). Participants were 15 psychology students (aged 18–30), 6 men and 9 women. There were between three to six participants in each of three climate cafés. Participants were assigned pseudonyms as Young person 1 (YP1), Young person 2 (YP2), etc., in transcription.

A guided discussion schedule was used to frame the discussion within the climate cafés. Participants were invited to bring an image that represents climate change for them. The use of visual materials such as a photograph or chosen image allows the researcher and participants to explore feeling and emotion through time and allows participants to be involved by choosing their own visual material, thus facilitating agency in the research process (Reavey & Johnson, 2017). Visual methods can be a way to disrupt well-rehearsed narratives on a topic (Reavey & Johnson, 2008). Following discussion of their images, participants were asked to imagine the climate in ten years' time, and in the future beyond that. They were then invited to imagine a possible "desired" climate future and what possible "self" one would have to be/ become to bring about that future. Each Climate Café lasted for approximately 1 hour.

Deductive thematic analysis as outlined by Braun and Clarke (2022) was used to analyse textual data. The analytic focus was on detailing imaginative processes including imagined positive or negative future orientation, spheres of experience (proximal or distal to self), the semiotic resources used within imaginative loops, and dimensions of plausibility-implausibility in imagined scenarios.

Results

Representation of Climate Change

Young people were asked to bring images of what climate change meant to them. In most cases, images represented distal spheres of experience, where climate change was perceived as at a geographic distance in Australia or the Arctic, or UN level activism (Fig. 7.1).

One young person chose an image representing the devastating consequences to society of severe flooding caused by climate change:

> …it sort of like captures how changes in climate systems have this effect where they totally isolate parts of the world and leave [areas] totally unliveable…Climate change is almost like squeezing out parts of the world, making it smaller and smaller until people are going to have to only occupy these ever-decreasing regions. So, we'll have more people and less space. I think this picture …it looks to me like the train is still running and…it's almost like a metaphor for what we're doing. Like, we

Fig. 7.1 Distal images of climate change

> know this is happening…we're degrading the Earth…resources are finite and we're still just churning through them. But we're not stopping. So, it's almost like, oh no, the train might flood, but it has to keep fucking going. It's like no one is going to be like, OK, maybe we shouldn't take the train anymore. It's just no, we'll just carry on taking the train, even though it'll probably drown us. (Young Person 4).

The metaphor of a train captures an imaginary future whereby the train keeps going. In spite of the observation "we know this is happening", there is an asymmetry in that we refuse to change the everyday routines of daily life. An imagined future whereby parts of the world are becoming unlivable is perceived as ineffective in motivating collective change, "we're not stopping" (Fig. 7.2).

Three images represented climate change as proximal experience, encountered close to the young person's daily life. YP2 explained that he comes "from a long line of farmers" and entitled his photo "the death of the small farmer". He has observed first-hand the changes wrought by climate through changing wetter seasons, changing livelihoods and resultant economic problems. For YP3, she lived in the inner city, and she has noticed "the rate of consumption and how we litter". She chose

Fig. 7.2 Proximal images of climate change (photographs by research participants of an image that represents climate change for them)

an image of a bin that said *"bruscar"* [Irish for litter] as she wanted it to reflect both her personal and Ireland's national consumption.

Imagining a Climate Future

Two imagined climate futures were evident in young people's talk (Fig. 7.3a and b). One imaginative loop is best conceived as depicting individualistic actions leading to small, graduate, positive change (Fig. 7.3a). This imagined future drew on semiotic resources such as recycling, composting, changing patterns of food consumption such as becoming vegetarian or vegan, taking public transport, cycling, and voting, closely tied to existing social norms. Young people noted:

> Like you can do little bits, ..if you take the time to carefully make sure it's [packaging] recycled in the proper way. (YP3)
> Small things happen here and there that are like a nice little boost and make you feel a bit hopeful. (YP11)
> They have one of those bins in [town] down by the beach where I'm from, where you can put your compost. [People] have definitely become more aware of the environment and what's going on. So it is moving in the right direction. It's just a bit slow. (YP14)

These imaginings represent concrete actions that are individualistic, familiar, plausible to achieve, based on existing technologies. They are close to everyday life rather than futuristic social or technological change. As such, they represent "thin" imaginative resources, bounded by what is, rather than anticipated solutions that may be "coming but not yet here". One participant imagined *"small incremental changes that you could do that could ultimately lead to larger collective change"*. In the positive orientations to the future, imagination is small-scale and modest. It may be that either participants have a lack of access to cultural resources to imagine more creative transformative change or alternatively, imagining beyond the familiar feels unsafe and uncomfortable, thus inhibiting imagination. The second imagined future was one of either a gradual or sudden eco collapse. One young person compared history to a march

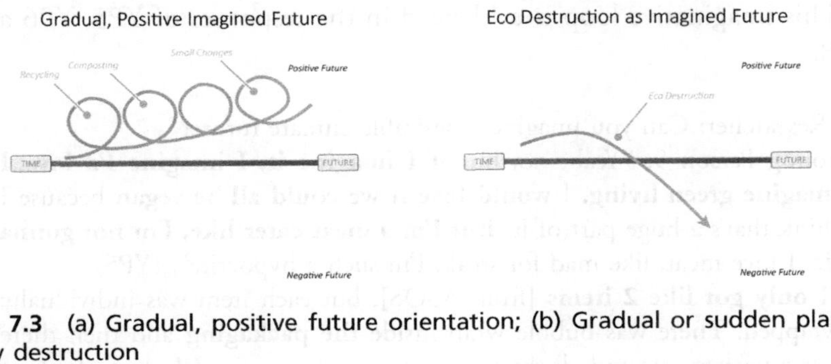

Fig. 7.3 (a) Gradual, positive future orientation; (b) Gradual or sudden planetary destruction

of progress in which humans spent ten thousand years going up a hill, getting to the top and then, "*we're kind of now going down*" '(YP2). Another spoke of a burnt world in which animals he had never seen but would love to see "*may no longer exist on the planet in the near future*", which "*is kind of scary*" (YP1).

Interestingly, in both positive and negative future-oriented imagining, these young people did not draw significantly on scientific research and evidence, indicating either a lack of comprehension of climate-related knowledge or a sense that it is not of direct relevance to their immediate or imagined lives.

Imagined Selves in a Desired Climate Future: Encountering Imaginative Boundaries

Climate café participants were invited to imagine a desired climate future and what possible "self" would have to be or become to bring about that future. While some participants mentioned concrete actions they are currently taking, such as 'reduce, reuse, recycle', a common pattern emerged whereby attempts at imaginings looped out to potential actions towards a positive future horizon. At this point, the imaginative loop became difficult to sustain and petered out as their imagination encountered semiotic borders (Picione & Valsiner, 2017), such as contradictions which emerged between their desired and real self, or between their limited agentive self and perceived uncontrollable corporate power (Fig. 7.4).

This imaginative loop is evidenced in the responses of YP5, YP6 and YP8.

> "Researcher: Can you imagine a desirable climate future?"
> Young Person 5: "Yeah, no, but **if I imagine it, I imagine I'd love..I imagine green living.** I **would love if we could all be vegan** because I think that's a huge part of it. **But I'm a meat eater like, I'm not gonna lie**, I love meat, like mad for steak, I'm such a hypocrite". (YP5)
> "I **only got like 2 items** [from ASOS], but each item was individually wrapped. There was bubble wrap inside **the packaging** and then there was a package around all the packaging in a box. . S**o, I have to look at**

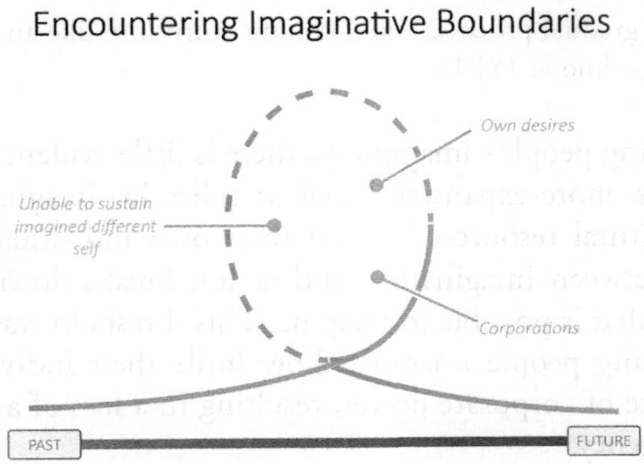

Fig. 7.4 Encountering imaginative boundaries

> **myself then and say,** well, is it something that I really care about or is it something that when it's pointed out to me, I notice it? It's not going to change the fact that **if ASOS has a good sale, I still buy off ASOS. The hard truth.** (YP6)
> "**I've been threatening to try and grow my own vegetables** for about, I'd say five years now, but that still hasn't happened so, I don't know, **that one's on the long finger**". (YP8)

In each case, attempts at imaginings reached a threshold whereby the imagined actions had undesired real-world consequences and the positive future-oriented possibilities opened up within the imaginative loop became difficult to sustain (Fig. 7.4). A moral and ethical imagining—I *should* change patterns of consumption/ How I *want* to be—came up against a resistance to change normative ways of being. The orientation to the future lacked persistence, became disconnected, and broke down as the loop returned to the 'here and now'. The imaginative contents indicates participants experienced a paucity of imaginative cultural resources, as explained by YP13:

> Uh, I think I'm really not sure. But I think for that to happen, I think we would have to have a lot of appropriate alternatives to what we're doing at the moment that are probably not in existence yet. I don't know. I'd say it

> will be like a gradual process. Like I couldn't turn into like an eco-warrior overnight, you know? (YP13)

In these young people's imaginings, there is little evidence of a linkage with a broader, more expansive social or collective imagination. In the absence of cultural resources beyond their own individual efforts, the relationship between imagination and action breaks down, reaching a threshold on what is possible to imagine. This threshold was also quickly reached as young people imagined how futile their individual actions were in the face of corporate power, resulting in a loss of a sense of self-efficacy and agency.

> I know I sound so pessimistic but it's like, for every small or big thing that's done for the climate, it seems like there's other stuff going in in the background that's just cancelling it out. Like all these data hubs and AI consuming loads of energy. I know we've got more green energy now in Ireland, but it's not matching what is being consumed. Like until the big boys at the top and the countries are on board, the ones who are producing the most amount of carbon emissions until they're on board, then we're kinda just like I don't know, destroying the planet and that's it. Pessimism! (YP8).

The capacity to believe in positive, gradual small changes evidenced in the earlier discussion above could not be imaginatively sustained in the face of global inequalities of power and resources. As one participant noted "*It's like we've discovered how to have a space hotel [on Mars] and I can't afford an Airbnb*" (YP8).

Imagining a Politics of Care

Terrains of despair, such as evidenced above, can be contrasted with terrains of care within imaginary loops. The strongest account of care-related discourse called upon a past experience of collective action that was successful, and this remembered experience then provided a semiotic resource for imagined future climate success similarly based on collective action. In this imaginative process, the imaginary loop first circles back to the past, before moving forward in an expansionary arc towards a positive climate future (Fig. 7.5).

This imaginative narrative was recounted by YP2:

> There's a huge mural in the village where I live and its #noincinerator; basically my entire community got together to protest this incinerator, managed to convince a big judge, managed to convince other powerful people this is bad- we don't want this, and they actually had a positive outcome..I would hope in the future in relation to climate change, communities would have a stronger sense of it if we work together, we could prevent this. (YP2)

Imagination oriented towards the past, such as imagining successful collective action to halt an incinerator being built in a local town, was then drawn upon as a cultural resource for a plausible vision of cooperation for climate action. This involves a shift from individual imaginaries to collective imagination, defined by Zittoun and Gillespie (2018) as "loops of imagination distributed among many people over time, with partly shared and crystallized outputs of imagination being the resources utilised in subsequent imaginings" (p. 20). Also, the authors assume "that collective imaginations can have large-scale outcomes, which not only affect one person, but a community or a

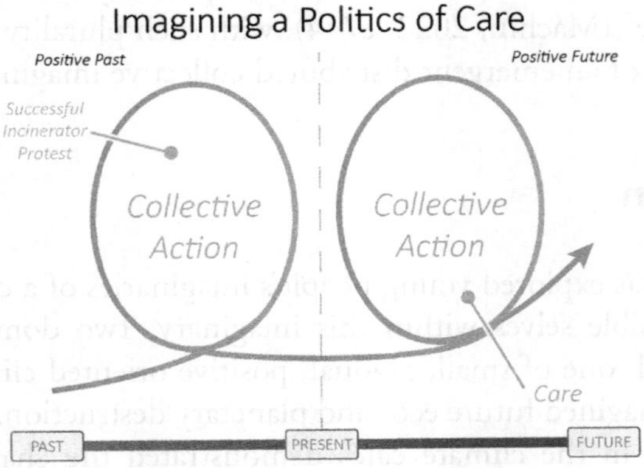

Fig. 7.5 Imagining a politics of care

society at large" (p. 20). This raises the possibility that, in imagination and in life, the small-scale, gradual changes made by individuals as outlined earlier, towards a positive future climate orientation could reach a "social tipping point" whereby apparently stable societal norms can be effectively overturned by the efforts of small but committed minorities (Centola et al., 2018). Centola and colleagues suggest that this social tipping point can be reached at a threshold of 25% of a community or group, that is, when one in four people enact a new social norm, it can then spread throughout the larger society. Iterative loops of imagination and action over time may become powerful mediators in facilitating such a tipping point.

Within this discussion, however, there is also evidence of a dubious imaginary that is critical of a climate-absent older generation. Imaginaries of care draw into the loop those perceived not to care; as one participant noted, "Do people of different generations not care in the same way as Greta Thunberg does, because they are not going to be here in 50 years? It is us that it is going to start affecting, and is already affecting" (YP1). This discussion within our climate café echoes Machin's (2022) notion of "sceptical" and "radical democratic imaginaries", whereby a sceptical imaginary is one which "casts doubt on the possibilities of democratic mechanisms to respond adequately and swiftly to the problem of climate change", and a radical democratic imaginary is "responsive to the inevitable and valuable plurality around the issue of climate change" (Machin, 2022, e744), with such plurality giving rise to the possibility of an emergent distributed collective imaginary.

Conclusion

This chapter has explored young people's imaginaries of a climate future and their possible selves within this imaginary. Two dominant imaginaries emerged, one of small, gradual, positive oriented climate change, the other of imagined future eco- and planetary destruction. Within this, the discussion in the climate cafés demonstrated the challenge young people imagined of introducing novel pro-ecological climate actions into their everyday routines while simultaneously wanting to carry on with their usual lives as if nothing dramatic is happening, in spite of an awareness of a changing climate. The challenge to sustain a positive imaginary self-in-climate trajectory into the future was linked to "thin" cultural resources drawn into the imaginative loop; most accessible were imaginings of individual level actions but once the energy consumption of corporations and big business became drawn into the imaginative loop, it resulted in the immobilisation of imagination. In contract, a politics of care mobilised imaginative resources and in imagination, individuals regained a sense of their significance. This has implications for education, the need for sustained, informed public dialogue, and mobilisation of individual and collective imaginations for community mobilisation as a tool towards a positive climate future.

References

Addressing climate change concerns in practice. (n.d.). Retrieved October 9, 2024, from https://www.apa.org/monitor/2021/03/ce-climate-change

Agency, E. P. (n.d.). *Ireland's state of the environment report 2024*. Retrieved October 9, 2024, from https://www.epa.ie/publications/monitoring--assessment/assessment/state-of-the-environment/irelands-state-of-the-environment-report-2024.php

Braun, V., & Clarke, V. (2022). *Thematic analysis: A practical guide*. Sage.

Clayton, S. (2020). Climate anxiety: Psychological responses to climate change. *Journal of Anxiety Disorders, 74*, Article 102263. https://doi.org/10.1016/j.janxdis.2020.102263

Centola, D., Becker, J., Brackbill, D., & Baronchelli, A. (2018). Experimental evidence for tipping points in social convention. *Science, 360*(6393), 1116–1119.

Demneh, M. T., & Darani, Z. H. (2020). From remembering to futuring: Preparing children for Anthropocene. *Journal of Environmental Studies and Sciences, 10*(4), 369–379. https://doi.org/10.1007/s13412-020-00634-5

Environmental Protection Agency, Ireland. (2024). *Climate change in the Irish mind*. Retrieved August 7, 2025, https://www.epa.ie/our-services/monitoring--assessment/climate-change/climate-change-in-the-irish-mind/

Flood, S., O'Higgins, T. G., & Lago, M. (2020). The promise and pitfalls of ecosystem services classification and valuation. In *Ecosystem-based management, ecosystem services and aquatic biodiversity: Theory, tools and applications* (pp. 87–103). Springer.

Galway, L. P., & Field, E. (2023). Climate emotions and anxiety among young people in Canada: A national survey and call to action. *The Journal of Climate Change and Health, 9*, Article 100204. https://doi.org/10.1016/j.joclim.2023.100204

Hajek, A., & König, H. H. (2022). Climate anxiety in Germany. *Public Health, 212*, 89–94. https://doi.org/10.1016/j.puhe.2022.09.007

Heeren, A., & Asmundson, G. J. G. (2023). Understanding climate anxiety: What decision-makers, health care providers, and the mental health community need to know to promote adaptative coping. *Journal of Anxiety Disorders, 93*, Article 102654. https://doi.org/10.1016/j.janxdis.2022.102654

Hickman, C., Marks, E., Pihkala, P., Clayton, S., Lewandowski, R. E., Mayall, E. E., Wray, B., Mellor, C., & van Susteren, L. (2021). Climate anxiety in children and young people and their beliefs about government responses to climate change: A global survey. *The Lancet Planetary Health, 5*(12), e863–e873. https://doi.org/10.1016/S2542-5196(21)00278-3

Hornsey, M. J., & Fielding, K. S. (2020). Understanding (and reducing) inaction on climate change. *Social Issues and Policy Review, 14*(1), 3–35. https://doi.org/10.1111/sipr.12058

Innocenti, M., Perilli, A., Santarelli, G., Carluccio, N., Zjalic, D., Acquadro Maran, D., Ciabini, L., & Cadeddu, C. (2023). How does climate change worry influence the relationship between climate change anxiety and eco-paralysis? A moderation study. *Climate, 11*(9), Article 9. https://doi.org/10.3390/cli11090190

IPCC (2021). *Summary for Policymakers. In: Climate Change 2021: The Physical Science Basis. Contribution of Working Group I to the Sixth Assessment Report of the Intergovernmental Panel on Climate Change*. Retrieved October 9, 2024, from https://researchspace.csir.co.za/items/0ba1dab4-637e-4929-9f64-442ce2b93bd1

Kelly, A. (2024). *An investigation of the psychological impacts of climate change on a sample of the Irish population.* Thesis, Institute of Art, Design + Technology. https://illustro-iadt.figshare.com/articles/thesis/An_investigation_of_the_psychological_impacts_of_climate_change_on_a_sample_of_the_Irish_population/25163747/1

Kricorian, K., & Turner, K. (2022). *Climate change and eco-anxiety in the US: Predictors, correlates, and potential solutions* (p. 2022.08.28.22279314). medRxiv. https://doi.org/10.1101/2022.08.28.22279314

Laino, E., & Iglesias, G. (2023). Extreme climate change hazards and impacts on European coastal cities: A review. *Renewable and Sustainable Energy Reviews, 184*, Article 113587. https://doi.org/10.1016/j.rser.2023.113587

Machin, A. (2022). Climates of democracy: Skeptical, rational, and radical imaginaries. *Wiley Interdisciplinary Reviews: Climate Change, 13*(4), Article e774.

Martin, G., Cosma, A., Roswell, T., Anderson, M., Treble, M., Leslie, K., Card, K. G., Closson, K., Kennedy, A., & Gislason, M. (2023). Measuring negative emotional responses to climate change among young people in survey research: A systematic review. *Social Science & Medicine, 329*, Article 116008. https://doi.org/10.1016/j.socscimed.2023.116008

Martin, G., Reilly, K., Everitt, H., & Gilliland, J. A. (2022). Review: The impact of climate change awareness on children's mental well-being and negative emotions—A scoping review. *Child and Adolescent Mental Health, 27*(1), 59–72. https://doi.org/10.1111/camh.12525

Masson-Delmotte, V., Zhai, P., Pirani, A., Connors, S. L., Péan, C., Berger, S., Caud, N., Chen, Y., Goldfarb, L., Gomis, M. I., Huang, M., Leitzell, K., Lonnoy, E., Matthews, J. B. R., Maycock, T. K., Waterfield, T., Yelekçi, Ö., Yu, R., & Zhou, B. (Eds.). (2021). *Climate change 2021: The physical science basis. Contribution of working group i to the sixth assessment report of the intergovernmental panel on climate change*. Cambridge University Press. https://doi.org/10.1017/9781009157896

Mulgan, G. (2022). *Another world is possible: How to reignite social and political imagination*. Hurst Publishers.

Picione, R. D. L., & Valsiner, J. (2017). Psychological functions of semiotic borders in sense-making: Liminality of narrative processes. *Europe's Journal of Psychology, 13*(3), 532.

Reavey, P., & Johnson, K. (2008). Visual methodologies: using and interpreting images in qualitative psychology. In *Handbook of qualitative research in psychology*. Sage

Reavey, P. & Johnson, K. (2017). Visual approaches: Using and interpreting images. *The SAGE handbook of qualitative research in psychology*, 354–373. https://doi.org/10.4135/9781526405555.n21.

Reyes, M. E. S., Carmen, B. P. B., Luminarias, M. E. P., Mangulabnan, S. A. N. B., & Ogunbode, C. A. (2023). An investigation into the relationship between climate change anxiety and mental health among Gen Z

Filipinos. *Current Psychology, 42*(9), 7448–7456. https://doi.org/10.1007/s12144-021-02099-3

Stanley, S. K., Hogg, T. L., Leviston, Z., & Walker, I. (2021). From anger to action: Differential impacts of eco-anxiety, eco-depression, and eco-anger on climate action and wellbeing. *The Journal of Climate Change and Health, 1*, Article 100003. https://doi.org/10.1016/j.joclim.2021.100003

Tripathy, K. P., Mukherjee, S., Mishra, A. K., Mann, M. E., & Williams, A. P. (2023). Climate change will accelerate the high-end risk of compound drought and heatwave events. *Proceedings of the National Academy of Sciences, 120*(28), Article e2219825120.

The nature, significance, and influence of perceived personal experience of climate change—Reser—2020—WIREs Climate Change—Wiley Online Library. (n.d.). Retrieved October 9, 2024, from https://wires.onlinelibrary.wiley.com/doi/abs/10.1002/wcc.668?casa_token=D9zYF4zYl5QAAAAA:LYE8fxn7SryQ_iEuQFRIUy1d3t5XAbjpP9osyw01JgfPCH_V9H9ub5hbCB7YqlmWLOCmszyNfzAp1g

Young People's Coping Strategies Concerning Climate Change: Relations to Perceived Communication With Parents and Friends and Proenvironmental Behavior—Maria Ojala, Hans Bengtsson, 2019. (n.d.). Retrieved October 9, 2024, from https://journals.sagepub.com/doi/full/10.1177/0013916518763894?casa_token=AWeKhlxqoWcAAAAA%3A1p8Z540qqx5aEelZXV7LA491Qg6H3x7qbpMf-JoIumWVYXEJHGYHGqcF9TFRgC8zcffp62YRxJo

Yusoff, K., & Gabrys, J. (2011). Climate change and the imagination. *Wiley Interdisciplinary Reviews: Climate Change, 2*(4), 516–534.

Vygotsky, L. (2004). Imagination and creativity in childhood. *Journal of Russian and East European Psychology, 42*(2), 7–97.

Zittoun, T. (2017). Symbolic resources and sense-making in learning and instruction. *European Journal of Psychology of Education, 32*(1), 1–20. https://doi.org/10.1007/s10212-016-0310-0

Zittoun, T., & Cerchia, F. (2013). Imagination as expansion of experience. *Integrative Psychological and Behavioral Science, 47*(3), 305–324. https://doi.org/10.1007/s12124-013-9234-2

Zittoun, T., & Gillespie, A. (2016a). Imagination: Creating alternatives in everyday life. In V. P. Glăveanu (Ed.), *The Palgrave handbook of creativity and culture research* (pp. 225–242). Palgrave Macmillan. https://doi.org/10.1057/978-1-137-46344-9_11

Zittoun, T. & Gillespie, A. (2016b). *Imagination in human and cultural development*. Routledge.

Zittoun, T., & Gillespie, A. (2018). Imagining the collective future: A sociocultural perspective. In C. de Saint-Laurent, S. Obradović, & K. R. Carriere (Eds.), *Imagining collective futures: Perspectives from social, cultural and political psychology* (pp. 15–37). Springer International Publishing. https://doi.org/10.1007/978-3-319-76051-3_2

8

Being Naturally Connected: The Role of Environmental Sensitivity in Building Resilience in the Climate Crisis

Alessandra Sperati, Maria Spinelli, Mirco Fasolo, and Francesca Lionetti

Introduction

The relationship between human and nature can be considered a real give-and-take in which both influence each other. The interest in understanding this bi-directional connection has increased over the last two decades likely due to the surging evidence on benefits of engaging with nature for human health (Chawla, 2020). The physical contact with nature, including direct exposures to natural settings, indeed positively

A. Sperati (✉)
Department of Neurosciences, Imaging and Clinical Sciences, University G. d'Annunzio, Chieti Pescara, Italy
e-mail: alessandra.sperati@unich.it

M. Spinelli · M. Fasolo
Department of Psychology, University G. d'Annunzio, Chieti-Pescara, Italy

F. Lionetti
Department of Brain and Behavioural Sciences, University of Pavia, Pavia, Italy

T. Skillington and A. Setti (eds.), *Climate Change Resilience Across Societal Contexts*, Energy, Climate and the Environment,
https://doi.org/10.1007/978-3-031-95020-9_8

impacts human daily lives and overall individual well-being (for review see Jimenez et al., 2021, Barragan-Jason et al., 2023).

Just as nature is protective for humans, humans are innately prone to be connected with nature (see the biophilia hypothesis, Wilson, 1984), showing a spontaneous attraction to nature since early in life and feeling themselves as part of the natural world, as broadly captured by the dimension of psychological connection to nature. According to empirical evidence, being psychologically connected to nature is related to pro-environmental behaviours and allows for nature conservation actions (Whitburn et al., 2020).

However, the human behaviours towards nature could be also destructive, threatening both the environmental sustainability and the human welfare. Moreover, though all humans are supposed to be psychologically connected with nature, recent evidence showed that some individual differences might account for the bi-directional flourishing between nature and humans (Dunne et al., 2024; Sumner et al., 2022) and thus, for fostering strategies and attitudes to address the climate crisis.

To understand what individual variables contribute to a heightened psychological connection to nature and pro-environmental attitudes, the individual trait of Environmental Sensitivity (ES, Pluess, 2015) has recently gained increasing attention, as among individuals who tend to exhibit a particularly strong connection to nature are those identified as highly environmentally sensitive (Black & Kern, 2020). At a phenotypical level, as captured and described by the Sensory Processing Sensitivity trait (SPS, Aron & Aron, 1997; Pluess et al., 2023), heightened levels of ES encompass an in-depth emotional and cognitive processing that is likely responsible of an increased sensitivity to stimuli, stronger emotional reactivity, and an elevated responsivity to environmental contexts, for better and for worse (i.e., as outlined by the Differential Susceptibility model in Belsky et al., 2007, described below). Importantly, individuals scoring high on ES are at greater risk for internalizing symptoms, such as anxiety and depression (Liss et al., 2005), and difficulties in emotion regulation, especially when the environment is less than optimal (Lionetti et al., 2022; Sperati et al., 2024). At the same time, they show stronger responsiveness to supportive and positive environments, experiencing greater benefits than less-sensitive individuals from prevention and intervention programs (Pluess & Boniwell, 2015; Ceccon et al., 2023; Nocentini et al., 2018).

This has implications for highly sensitive people's responses to positive effect of nature, as well as to the adverse effects of environmental crises (e.g., a greater risk of eco-anxiety), and for fostering pro environmental behaviours for facing the climate crisis. Specifically, across adult and child population, around 30% of individuals have been found being high on this trait (Pluess et al., 2018; Lionetti et al., 2018). Highly sensitive individuals, due to their deeper processing of details and heightened appreciation of positive aspects in their surroundings (Aron 2002; Chou, 2024; Jagiellowicz et al., 2011; De Gucht et al., 2022), are likely to form stronger connections with green environments and, consequently, derive greater benefits from the power of nature. Similarly, thanks to their increased disposition to be emotionally attuned with others (i.e., greater empathy) (Acevedo et al., 2014; 2021) they are more likely to be psychologically connected to nature, including being more emotionally in contact with animals and plants, and thus to involve more in protective and pro-environmental actions (Dunne et al., 2024). Establishing a contact with nature, especially early in life, and fostering the potentially innate connection of highly sensitive individuals to the natural environment could contribute to building resilience and provide a sense of agency when facing the climate crisis. The fact that adults with higher scores on sensitivity also report more pro-environmental behaviours is encouraging and this could potentially combat negative emotions generated by experiencing the changes in natural surroundings associated with climate change. Also, nature has been found as a factor promoting well-being in highly sensitive people (Setti & Intyre, 2023). Overall, this suggests that highly sensitive individuals can give to nature, and nature can give back to them.

The research interest in investigating the role of individual differences in ES in response to green environment is increasing. Available literature on this topic includes a limited number of studies and currently the most are focused on adults (Setti et al., 2022). A theoretical reflection and a qualitative review of extant evidence on this nascent field can inspire about implications for both research and policy fields. In this book chapter, we will first provide an overview of the topic of human-nature bi-directional relationship, focusing on the way through which the nature is important to human and vice versa (i.e., the bi-directional flourishing). We then introduce the Environmental Sensitivity framework followed by the theoretical rationale and significance of considering the ES trait as a pertinent concept in the realm of green studies. Finally, we will qualitatively review the available evidence involving adult populations, discussing the significant potential of this emerging line of research for practical applications and highlighting future directions, including the relevance of exploring this association early in childhood.

The Bi-Directional Flourishing Between Human and Nature

Nature Gives to Human: The Effects of Physical Contact with Green Environments

It is widely acknowledged that taking a walk in the wild as well as playing in a park is good for human health. A large body of research across several scientific disciplines (e.g., from biology to psychology) has provided substantial evidence on the in-depth physical and psychological effects of being exposed to green and natural environments in both children and adults (for review, see Barragan-Jason et al., 2023; Jimenez et al., 2021; Song et al., 2016). Specifically, meta-analytic findings on adults showed the green experience linked with several neurophysiological benefits. For instance, an empirical study found that people who were exposed to real plants showed greater activity in the pre-frontal cortex, an area mostly implicated in the emotion regulation processes, compared to those who were exposed to images of the same plants (Igarashi et al., 2015). Other physiological benefits of nature exposure were also found in relation to the decrease in blood pressure, improvement in the immune function, and reduction of cortisol levels in people when undertaking a trip in the forest (see review of empirical findings in Jimenez et al., 2021). The experience of green environments has been also linked to mental health, with positive effects on both emotional and cognitive skills (e.g., Bratman et al., 2015; for a review Ohly et al., 2016). Findings suggested that adults who experienced a natural environment (e.g., walking in the wild) were the same who reported a decrease in depression and anxiety together with a preservation of positive affect and greater working memory performance than those who were exposed to a non-green environment (i.e., urban context). Similar benefits have been found in children samples. Precisely, review of empirical findings showed that having green spaces in their surrounding (i.e., home, neighbourhoods, schools) impacts children cognitive performance (Chawla, 2020), overall suggesting a relevant role of outdoor environments in the brain development. Moreover, access to different types of green spaces is related to improved physical (e.g., lower risk for obesity and for vitamin D deficiency) and psychological well-being in children, such as improving attention focus, building confidence, and forming supportive peer relationships, along with less socio-emotional difficulties and decrease of psychological distress (for a review, see McCormick, 2017). Taking together, these findings suggest that spending time in natural environments allows both adults and children to overcome physical and mental fatigue across several mechanisms that seem to involve different levels of the human experience. According to the theoretical reasoning in this field, this protective power of nature for humans could be because the exposure to green activates the parasympathetic nervous system that in turn is responsible for reducing stress and autonomic arousal (see the Stress Reduction theory, Urlich, 1984). Similarly, natural environment allows restoring the capacity to direct human attention that could be threaten by the modern life styles (e.g., frenetic daily life) (see the attention restoration theory, Kaplan, 1995). Interestingly, this thinking assumes that the power of nature is accessible due to humans' spontaneous connection to natural realm and that feeling psychologically connected to the natural world facilitates the potential natural benefits to accrue, leading to the concept of psychological connection with nature. Hence, it is reasonable to hypothesize that for highly sensitive individuals, at greater risk of emotional overwhelming, the contact with nature as well as green interventions represent significant opportunities for increasing their well-being, allowing them to flourish, by benefiting at most from this 'green power' (Setti & Intyre, 2023).

Humans Matter for Nature: The Role of Psychological Connection to Nature

Looking at a young child in a natural environment, one might be captured by children's spontaneous approach to natural elements, including pleasure in sensory details of plants and flowers as well as a loving care behaviour towards animals and other species lives (Chawla, 2020). This makes evident the children curiosity and fascination with other living beings and suggests an innate kinship among humans with nature (see the biophilia hypothesis, Wilson, 1984). This human tendency to nature has been further supported by empirical studies showing children to be emotionally attracted and cognitively captured by nature since preschool years (for review see Chawla, 2020). Studies involving adults provided similar supportive findings, suggesting a human preference for savannah-like landscapes and a shared idea about the restorative function of being immerged in nature (Joye & Dewitte, 2018). This "being prone" to nature has been integrated into the broader concept of psychological connection with nature (Mayer & Frantz, 2004). Though operationalized in various terms and across several measures, this psychological nature connectedness could be broadly defined as an individual attitude towards nature involving affective, cognitive, and behavioural dimensions. Overall, it encompasses an in-depth human relatedness to nature, comprising the emotional attraction and affiliation with nature, a sense of love and caring towards nature, a curiosity about natural phenomena up to the inclusion of nature in self, and the cognitive understanding of human-nature interdependency (see Tam, 2013, Barragan-Jason et al., 2023, for reviews).

The psychological connection to nature is interrelated with nature contact, as further supported by empirical findings. Specifically, people with greater psychological connection to nature are more likely to seek out more nature contact and to spend more time in natural settings than indoor. Furthermore, when exposed to green environment, people with high nature psychological connectedness were the same reporting a state of greater awareness of nature and empathy for plants and animals at the moment, which in turn impacts their sense of well-being (Mayer et al., 2009; Martin et al., 2020).

Besides enabling humans to benefit from the power of nature, in a bi-directional way, the psychological connection is also related to a heightened motivation to protect nature, enhancing one's engagement in pro-environmental behaviours (see meta-analytic findings, Whitburn et al., 2020; Martin et al., 2020). Pro-environmental behaviours broadly referred to all that actions contributing to environmental protection and conservation (Axelrod & Lehman, 1993) that include both household actions, such as recycling and buying ecological products, and nature conservation behaviours, for example, volunteering in nature organization for helping the environment (Martin et al., 2020). However, though such a spontaneous kinship to nature, the relationship between human and nature could be sometimes threatened, as discussed below. This seems to suggest that other variables could count in this regard. Interestingly, recent studies (Dunne et al., 2024) propose the ES individual trait as a protective factor for pro-environmental behaviours, with those high on this trait potentially more interested in engaging in caring behaviours towards the environment. Notably, given the greater involvement with nature, highly sensitive people may also be more likely to contribute to actions addressing the climate crisis and the challenging feelings associated with it (e.g., eco-anxiety).

Existing Threats to the Bi-Directional Flourishing Between Human and Nature

There is growing evidence that individual contact with nature is decreasing around the world. People are spending less time outside and more time indoor. Urban areas are becoming more densely developed, and frenetic lifestyles often exclude "green-time" moments. This suggests a decline in opportunities to experience nature, along with a waning interest and motivation towards nature-related activities, making it challenging for adults to pass this connection on to children (Chawla, 2020).

Besides this nature disengagement, we are currently facing the environmental anthropogenic degradation that is one of the biggest human challenges of the twenty-first century. Human behaviours (e.g., habitat destruction, pollution, overharvesting) are threatening the nature ecosystem conservation, compromising the environmental quality and positing serious issues to human health (WHO, 2018). Both the disengagement from green environments and the existence of destructive behaviours towards nature represent a proper climate crisis, calling for further research in psychology field to understand what variables account for promoting a flourishing human-nature relationship.

A deeper psychological connection with nature has been shown to partially explain why some people engage more with nature and behave more pro-environmentally than others (see meta-analytic findings in Whitburn et al., 2020). Among other individual variables that might contribute to a stronger human-nature relationship, the Environmental Sensitivity individual trait (ES, Pluess, 2015) has recently gained increasing attention. In the following paragraph, we introduce the Environmental Sensitivity meta-framework and the rationale for which it would be plausible to hypothesize its role in the green world.

Environmental Sensitivity: What is About

Environmental Sensitivity as a Meta-Framework

To survive, both animals and humans are programmed to perceive and respond to specific physical and social stimuli in the surrounding environment but, interestingly, substantial inter-individual differences in sensitivity to the environment have been found. The Environmental Sensitivity refers to a basic trait—with genetic, neurophysiological, and behavioural correlates—that captures and explains these individual differences in perceiving, processing, and responding to internal and external stimuli, both positive and negative ones (Pluess, 2015). Such individual differences in responding to the environment have been found in most organisms across species including humans (for review, see Wolf et al., 2008). Across children, adolescents, and adults, a continuum from low to high sensitivity has been observed, with a majority (around 40% of the general population) showing medium levels of sensitivity, and two sizable minorities characterized by particularly high (31%) and low

(~29%) sensitivity (Lionetti et al., 2018). Environmentally highly sensitive individuals, often referred to as "orchid" people (as illustrated by the floral metaphor reported in Ellis & Boyce, 2011), tend to respond more negatively to adverse environments but also more positively to nurturing ones (for a comprehensive review, see Greven et al., 2019).

Theoretical models addressing sensitivity to both positive and negative contexts are included into the Environmental Sensitivity meta-framework. This framework integrates three parallel lines of research that approach sensitivity from evolutionary, developmental, and personality perspective, namely: the Differential Susceptibility model (Belsky, 1997; Belsky et al., 2007; Belsky & Pluess, 2009), the Biological Sensitivity to context model (Boyce & Ellis, 2005), and Sensory Processing Sensitivity theory (Aron & Aron, 1997) (see Fig. 8.1). More in details, the Differential Susceptibility Model (Belsky, 1997; Belsky et al., 2007; Belsky & Pluess, 2009), based on an evolutionary perspective, suggests that individual vulnerability characteristics (e.g., difficult temperament, high physiological reactivity) may actually reflect a general sensitivity to both harsh and supportive exposures, making some children more susceptible than others "for better and for worse" (Belsky, 2007).

From a developmental trajectories perspective, more susceptible individuals are more likely to experience negative outcomes in adverse environments (i.e., "for worse") as well as positive outcomes in supportive environments (i.e., "for better"). In a similar but parallel line of theoretical reasoning, the Biological Sensitivity to Context model was developed (Boyce et al., 1995; Boyce & Ellis 2005). This model suggests that differences in sensitivity reflect differences in physiological reactivity that are shaped by early experiences. In a nutshell, in line with the notion of conditional adaptation to stress (Ellis et al., 2011), individuals would develop a heightened physiological reactivity (e.g., arterial pressure, cortisol reactivity) shaped by early challenging environmental influences, increasing their capacity to respond to threats and enabling them to benefit greatly from environmental resources. Finally, developed in the personality psychology field, the Sensory Processing Sensitivity theory (Aron & Aron, 1997) proposes a phenotypical perspective of sensitivity. It posits that the core mechanism of heightened sensitivity would be an in-depth processing of inputs that contributes to stronger

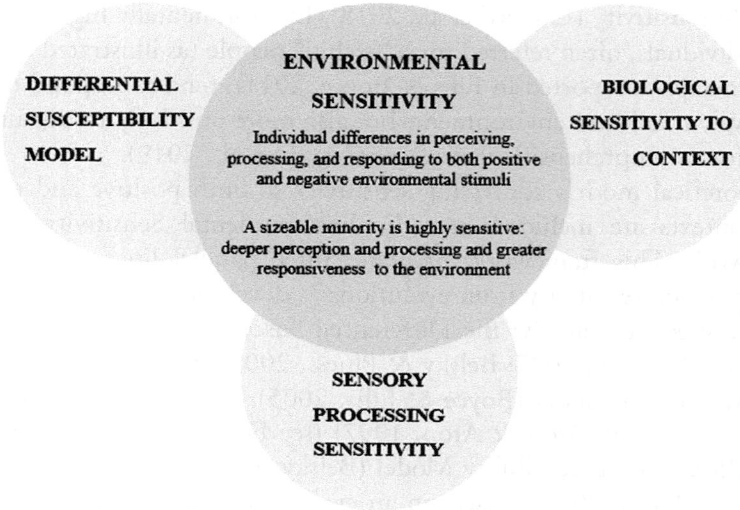

Fig. 8.1 Graphical representation of the environmental sensitivity meta-framework

emotional responses to the surrounding environment (Lionetti et al., 2018; Pluess et al., 2023). Though each of these models address the study of the person-by-environment interaction by yielding a unique contribution, all agree that people differ in their sensitivity for better and for worse, with some showing a heightened sensitivity to both positive and negative effects of the environment. More recently, the Vantage Sensitivity model (Pluess & Belsky, 2013) has been proposed to capture the bright side of sensitivity. In contrast to the diathesis-stress effect (i.e., suffering more when confronted with adversity due to individual vulnerability), this model suggests that some individuals are more responsive than others to enriched environments but those attributes that render some people more sensitive to the benefits of supportive experiences do not make them necessarily also more vulnerable to the negative effects of adverse conditions. The vantage sensitivity perspective proposes an adequate theoretical model to capture individual variation in response to exclusively positive environments as a function of individual characteristics, potentially including the individual responsiveness to green exposure with a greater benefit from it.

Environmental Sensitivity as an Individual Trait

Heightened Environmental Sensitivity is reflected in several markers at different levels of analysis (i.e., genetic, physiological, and behavioural) (for a comprehensive review, see Belsky & Pluess, 2009; Obradovic & Boyce, 2009). In the context of this chapter, we will particularly focus on a phenotypical perspective of sensitivity as proposed by the Sensory Processing Sensitivity psychological trait (SPS; Aron & Aron, 1997), given its relevance for understanding the human-nature bi-directional relationship.

According to SPS, heightened sensitivity seems to arise from a more reactive nervous system that implies a deeper processing of both cognitive and emotional stimuli (Acevedo et al., 2014, 2021) and can result in a stronger emotional and sensory reactivity. At the behavioural level, both theoretical thinking and empirical studies on SPS have shown that heightened sensitivity is reflected in a constellation of attributes, including depth of processing, tendency to overstimulation, increased emotional reactivity, and sensitive to subtleties (see also DOES acronym in Aron et al., 2012) (see Fig. 8.2). In other words, highly sensitive individuals are more likely to get easily overwhelmed, but, importantly, they also showed higher sensory pleasure and greater appreciation of positive subtleties in the surrounding (i.e., aesthetic sensitivity), likely due to their deeper processing that promotes greater awareness of environmental stimuli (De Gucht et al., 2022). Noteworthy, heightened SPS is also associated with empathy, with highly sensitive individuals found to have greater emotional attunement (Schaefer et al., 2022) and higher awareness of others' emotional states (Li et al., 2021).

Drawing on the SPS psychological trait, a range of measures, including questionnaires and observational rating systems, have been developed. Self-report questionnaire is largely used and easy to adopt. The first psychological measure to assess environmental sensitivity as reflected

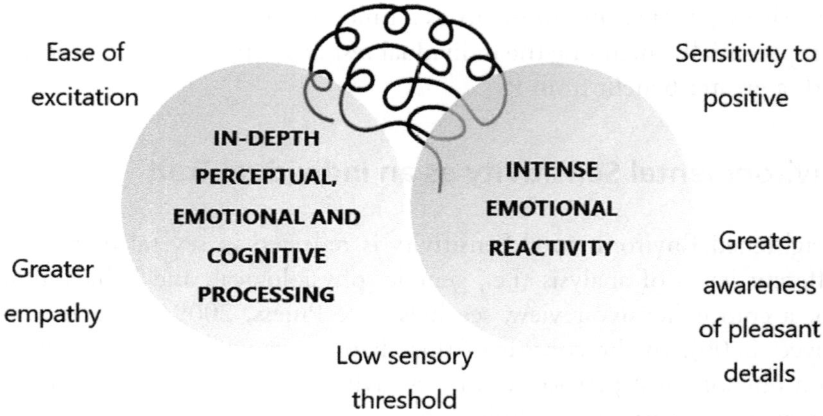

Fig. 8.2 Characteristics of high environmental sensitivity as captured by the notion of sensory processing sensitivity trait

in the SPS trait was the Highly Sensitive Person scale (HSP; Aron & Aron, 1997). This 27-item scale was originally developed based on series of qualitative interviews with individuals who self-described as "highly sensitive" (Aron & Aron, 1997), and then adapted and validated into a more recent 12-item self-report version for adults (HSP-12; Pluess et al., 2023; Lionetti et al., 2024). Transitioning to the assessment of SPS in children, the first questionnaire for measuring parent-reported sensitivity in children aged 3 to 16 years was proposed in The Highly Sensitive Child book (HSC; Aron, 2002). Since then, several measures have been proposed and validated to assess sensitivity, including both self-report (Pluess et al., 2018; Weyn et al., 2021) and parent-report questionnaires (Slagt et al., 2018, Sperati et al., 2022), as well as interviews (Kahkonen et al., 2024) and observational rating systems (Davies et al., 2021; Lionetti et al., 2019; Moscardino et al., 2021).

Why Should Environmental Sensitivity Matter for Human-Nature Relation?

Efforts and attempts to improve green experiences and promote pro-environmental behaviours have been written in the human agenda.

However, individual differences in psychological factors have been found to influence one's positive attitude and behaviours towards the environment. Understanding the characteristics of individuals who are more prone to act in environmentally responsible ways, strengthening pro environmental actions, could inform policies and intervention aimed at achieving environmental goals.

In this regard, heightened environmental sensitivity may make some individuals more likely to benefit from nature and to feel a stronger connection with nature and animals (Setti et al., 2022). Highly sensitive individuals, driven by their heightened appreciation for subtle positive aspects of their surroundings, are supposed to engage more with green environments, increasing their chances for benefiting from the power of nature. Indeed, since nature has repetitively been shown to have positive effects on mental health, including reducing anxiety and depression-related symptoms such as rumination, it may be especially precious for wellbeing in people high in ES who are more prone to internalizing symptoms, maladaptive emotional strategies, and negative feelings (Liss et al., 2005; Lionetti et al., 2022). Similarly, thanks to their higher emotional attunement with others (i.e., greater empathy competences), they are believed to be more psychologically connected with nature, including animals. For instance, they may capture more deeply stress of animals, becoming more emotionally connected with them and, as a result, more likely to take actions to protect them. Furthermore, highly sensitive people are more likely to adopt pro-environmental behaviours. Their heightened sensitivity and indepth processing may encourage them to pause before acting (see also the "pause-to-check" mechanism in highly sensitive children; Aron & Aron, 1997; Aron, 2002), allowing them time to reflect on the consequences of their actions, particularly those affecting the environment.

To date only a few empirical studies have investigated this hypothesis, as described below.

Empirical Evidence on the Role of Environmental Sensitivity in the Green Realm

A first insight into the fact that heightened environmental sensitivity may make individuals more involved with natural world came from a qualitative exploration of well-being in highly sensitive people (Black & Kern, 2020). According to the interviewees, half of the individuals high in ES reported that connecting with nature was a key factor in their well-being, emphasizing the restorative power of being among green. For instance, participants told how enjoyable it was walking through a nearby bush track, look up at the trees, and being mindful (Black and Kern, 2020). This finding is not surprising when considering that highly sensitive people are characterized by an increased sensitivity to aesthetically pleasing stimuli that allow them to appreciate works of art, music and harmonies, and beauty of nature (Aron & Aron, 1997; Aron et al., 2012).

Drawing from this qualitative suggestion, a multi-study paper investigated the hypothesized association between high ES and connectedness with nature and animals across two independent adult samples (Setti et al., 2022). In Study 1, 241 Irish adults were invited to fill out an online survey. Participants reported on their ES by completing the 27-item Highly Sensitive Person scale (Aron & Aron, 1997) capturing high sensitivity characteristics, such as ease of excitation, low sensory threshold, depth of processing and aesthetic sensitivity. The connectedness with nature was captured by items assessing an overall sense of belonging to natural world (e.g., "I often feel a sense of oneness with the natural world around me/I recognise and appreciate the intelligence of other living organisms"), while their connection to animals was indexed by the levels of affinity to pets (e.g., I love pets/My pet makes me feel happy/I consider a pet to be a friend) (Mayer & Frantz, 2004). Findings showed that being high on ES trait significantly predicted greater connection to nature but, noteworthy, the same support was not found for the connection with animals. Thus, findings were replicated in a second study (see Setti et al., 2022) including a sample of 144 Irish adults to consolidate significant results from Study 1 and to further explore the association with animal kinship by using a different measure previously adopted in American context (Herzog et al., 2015). Moving beyond the relation with pets, items included in the new scale captured a broader and deeper attitude towards human relation with animals (e.g., It is morally wrong to hunt wild animals just for sport/I sometimes get upset when I see wild animals in cages at zoos). Interestingly, when including the animal welfare and affinity to animal dimensions, the higher the environmental sensitivity, the greater involvement in animal lives, suggesting that proneness to feel in contact with animals is regardless of being in a specific relation to pets.

Overall, these pioneering studies suggest that individual differences in ES may explain different attitudes towards green realm, including the adoption of actions to protect both nature and animals, shedding light on heightened ES as a valuable strength for human-nature bi-directional flourishing. According to this, it is plausible to hypothesize that people high in ES are more likely to behave in a pro-environmentally way, paying more attention to the consequences of their behaviours, contributing to foster community resilience toward the climate changes. And this means that, for example, in schools, highly sensitive children may positively influence their peers through peer-to-peer programs.

In exploring empirically whether people high in sensitivity are more prone to adopt actions to protect the environment, a very recent study (Dunne et al., 2024) shed light on the potential underlying mechanisms implied in the broad relationship between heightened sensitivity and connectedness with nature. The first evidence in this regard found from the study was about the role of the awe (i.e., feeling smaller in front of the magnificence of the natural world) in mediating this association. Specifically, findings from a relatively large sample of 807 Irish adults surveyed online showed that the greater appreciation for positive aspects characterizing highly sensitive people seems to be responsible for their disposition to experience awe when exposed to the green realm that promotes a more intense connection with nature. In turn, this greater nature connectedness seems to motivate people, especially those high in ES, to take care of the environment and to act sustainably. In other words, it seems that, likely due to their feeling of wonder towards greatness of world, highly sensitive people wish to protect the great natural world they feel part. This process promotes positive active behaviours that, moving beyond green values or environmental concern, allow for most significant outcomes in the environment and contribute to strengthen human agency in protecting nature.

Overall, evidence suggests that people characterized by higher levels of ES are more likely to engage in concrete actions that account for environment protection, including both home-based attentions towards the environment (e.g., turning lights or water tap off) and broad pro-environmental volunteering actions (e.g., active promotion of attitudes for tackling climate change or for cleaning beaches or parks).

However, available studies are cross-sectional and limited to the adult population, overlooking the developmental aspect. Moreover, the most studies have a correlational design and data are based on self-report questionnaires. Further research is called to explore this relationship in childhood, since preschool and school years, when children begin to get aware of other emotional states and potentially towards the natural world too. Investigating the protective role of nature may offer insights for programs aimed at promoting children's well-being, especially for those high in ES who are more prone to externalizing and internalizing problems but also benefit more from enriched and positive environment, with nature being one such context. If the preliminary findings from adult studies, which suggest that highly sensitive adults tend to be particularly connected and attuned to nature, are replicated in child samples, this could provide new guidelines for peer-to-peer programs in educational settings. These programs could leverage sensitive children's heightened sensitivity to positive environmental details as a means to foster pro-environmental behaviours. Highly sensitive children could, in turn, become "leaders of positive contagion", serving as role models for their less-sensitive peers to encourage a protective attitude towards the environment.

Implications for Future Research

The identification of individual variables related to sensitivity and responsiveness to the environment adds to the understanding of the relationship between human and nature, with relevance for both individual well-being and environment.

On the individual level, contact with natural contexts and animals can be a potential source of prevention and recovery from stress, especially for highly sensitive people who are more prone to challenges in the emotional experience (Lionetti & Pluess, 2024), but also benefit significantly from enriched experiences. Similarly, nature-based or pet-based therapies could be potentially effective in such population. Further research focusing on the effect of nature on human well-being, depending on individual differences in ES, will provide empirical support to confirm that.

At the community level, knowledge about individual differences in ES and its relationship with the natural world can have implications for urban design and policymaking related to city liveability (Bratman et al., 2019). For example, awareness on individual variability in sensory stimulation could guide policies in offering services suitable for everyone, particularly benefiting those more prone to sensory overwhelm. This might include creating green spaces within cities that are easily accessible to people coming from different neighbourhoods. Furthermore, community actions should aim to address the climate injustice, such as growing up in nature-deprived contexts or highly polluted environments. This can have a significant impact on highly sensitive individuals, especially early in life. For instance, higher sensitive children may be more vulnerable to adverse environmental conditions, including growing up in highly urbanised areas with limited access to nature, especially if combined with other risk factors, e.g. lower socio-economic status. While this is true for all children, those higher in sensitivity may be at greater risk of such adverse effects. Conversely, children high in sensitivity may thrive in nature-rich environments, where they can build nature connectedness, ultimately becoming a resource for others' green awareness.

Finally, on the natural environment side, drawing from theoretical insights and empirical evidence on the association between environmental sensitivity and nature herein discussed, it seems that considering individual differences in ES might be of relevance for the growing urgency of the climate change and the environmental crisis (Watts et al., 2021). Individuals with high SPS, due to their sensitivity and deep connection to the welfare of nature and animals, though may be more vulnerable to suffer from the destruction of nature, i.e., eco-grief and eco-anxiety (Cunsolo et al., 2020; Chawla, 2020), conversely, they may also become promoters of ecological and sustainable behaviours, serving as ambassadors for climate change. Their greater awareness and caring actions towards the environment could model others' attention to nature and motivate them to protect it.

In conclusion, understanding the role of environmental sensitivity in the relationship between humans and nature is critical for enhancing individual well-being and addressing pressing environmental concerns. By identifying individual differences in how people respond to natural stimuli, we can inform interventions at both the individual and societal levels, for example better tailoring intervention programs, creating more liveable environments, and promoting ecological awareness and fostering more environmentally conscious communities.

References

Acevedo, B. P., Aron, E. N., Aron, A., Sangster, M. D., Collins, N., & Brown, L. L. (2014). The highly sensitive brain: an fMRI study of sensory processing sensitivity and response to others' emotions. *Brain and behavior, 4*(4), 580–594.

Acevedo, B. P., Santander, T., Marhenke, R., Aron, A., & Aron, E. (2021). Sensory processing sensitivity predicts individual differences in resting-state functional connectivity associated with depth of processing. *Neuropsychobiology, 80*(2), 185–200.

Aron, E. N., & Aron, A. (1997). Sensory-processing sensitivity and its relation to introversion and emotionality. *Journal of personality and social psychology, 73*(2), 345.

Aron, E. N. (2002). *The highly sensitive child. Helping our children thrive when the world overwhelms them.* Broadway Books.

Aron, E. N., Aron, A., & Jagiellowicz, J. (2012). Sensory processing sensitivity: A review in the light of the evolution of biological responsivity. *Personality and Social Psychology Review, 16*(3), 262–282.

Axelrod, L. J., & Lehman, D. R. (1993). Responding to environmental concerns: What factors guide individual action? *Journal of Environmental Psychology, 13*(2), 149–159.

Barragan-Jason, G., Loreau, M., de Mazancourt, C., Singer, M. C., & Parmesan, C. (2023). Psychological and physical connections with nature improve both human well-being and nature conservation: A systematic review of meta-analyses. *Biological Conservation, 277*, Article 109842.

Belsky, J. (1997). Variation in susceptibility to environmental influence: An evolutionary argument. *Psychological inquiry, 8*(3), 182–186.

Belsky, J., Bakermans-Kranenburg, M. J., & Van IJzendoorn, M. H. (2007). For better and for worse: Differential susceptibility to environmental influences. *Current Directions in Psychological Science, 16*(6), 300–304.

Belsky, J., & Pluess, M. (2009). Beyond diathesis stress: differential susceptibility to environmental influences. *Psychological Bulletin, 135*(6), 885.

Black, B. A., & Kern, M. L. (2020). A qualitative exploration of individual differences in wellbeing for highly sensitive individuals. *Palgrave Communications, 6*(1), 1–11.

Bratman, G. N., Daily, G. C., Levy, B. J., & Gross, J. J. (2015). The benefits of nature experience: Improved affect and cognition. *Landscape and Urban Planning, 138*, 41–50.

Bratman, G. N., Anderson, C. B., Berman, M. G., Cochran, B., De Vries, S., Flanders, J., ... & Daily, G. C. (2019). Nature and mental health: An ecosystem service perspective. *Science advances, 5*(7), eaax0903.

Boyce, W. T., Chesney, M., Alkon, A., Tschann, J. M., Adams, S., Chesterman, B., & Wara, D. (1995). Psychobiologic reactivity to stress and childhood respiratory illnesses: Results of two prospective studies. *Psychosomatic Medicine, 57*(5), 411–422.

Boyce, W. T., & Ellis, B. J. (2005). Biological sensitivity to context: I. An evolutionary–developmental theory of the origins and functions of stress reactivity. *Development and Psychopathology, 17*(2), 271–301.

Ceccon, C., Schachner, M. K., Lionetti, F., Pastore, M., Umaña-Taylor, A. J., & Moscardino, U. (2023). Efficacy of a cultural adaptation of the Identity Project intervention among adolescents attending multiethnic classrooms in Italy: A randomized controlled trial. *Child Development, 94*(5), 1162–1180.

Chawla, L. (2020). Childhood nature connection and constructive hope: A review of research on connecting with nature and coping with environmental loss. *People and Nature, 2*(3), 619–642.

Chou, Y. Y. P. (2024). Must one take the bitter with the sweet? Sensory processing sensitivity and factors associated with resilience among artistically inclined individuals. *Current Psychology, 43*(20), 18631–18646.

Cunsolo, A., Harper, S. L., Minor, K., Hayes, K., Williams, K. G., & Howard, C. (2020). Ecological grief and anxiety: the start of a healthy response to climate change? *The Lancet Planetary Health, 4*(7), e261–e263.

Davies, P. T., Hentges, R. F., Coe, J. L., Parry, L. Q., & Sturge-Apple, M. L. (2021). Children's dove temperament as a differential susceptibility factor in child rearing contexts. *Developmental Psychology, 57*(8), 1274.

De Gucht, V., Woestenburg, D. H., & Wilderjans, T. F. (2022). The different faces of (high) sensitivity, toward a more comprehensive measurement instrument. Development and validation of the sensory processing sensitivity questionnaire (SPSQ). *Journal of Personality Assessment, 104*(6), 784–799.

Dunne, H., Lionetti, F., Pluess, M., & Setti, A. (2024). Individual traits are associated with pro-environmental behaviour: Environmental sensitivity, nature connectedness and consideration for future consequences. *People and Nature, 6*(2), 586–597.

Ellis, B. J., & Boyce, W. T. (2011). Differential susceptibility to the environment: Toward an understanding of sensitivity to developmental experiences and context. *Development and Psychopathology, 23*, 1–5. https://doi.org/10.1017/S095457941000060X

Ellis, B. J., Boyce, W. T., Belsky, J., Bakermans-Kranenburg, M. J., & Van Ijzendoorn, M. H. (2011). Differential susceptibility to the environment: An evolutionary–neurodevelopmental theory. *Development and psychopathology, 23*(1), 7–28.

Greven, C. U., Lionetti, F., Booth, C., Aron, E. N., Fox, E., Schendan, H. E., & Homberg, J. (2019). Sensory processing sensitivity in the context of environmental sensitivity: A critical review and development of research agenda. *Neuroscience & Biobehavioral Reviews, 98*, 287–305.

Herzog, H., Grayson, S., & McCord, D. (2015). Brief measures of the animal attitude scale. *Anthrozoös, 28*(1), 145–152.

Igarashi, M., Miwa, M., Ikei, H., Song, C., Takagaki, M., & Miyazaki, Y. (2015). Physiological and psychological effects of viewing a kiwifruit (Actinidia deliciosa 'hayward') orchard landscape in summer in Japan.

International journal of environmental research and public health, 12(6), 6657–6668.

Jagiellowicz, J., Xu, X., Aron, A., Aron, E., Cao, G., Feng, T., & Weng, X. (2011). The trait of sensory processing sensitivity and neural responses to changes in visual scenes. *Social Cognitive and Affective Neuroscience, 6*(1), 38–47.

Jimenez, M. P., DeVille, N. V., Elliott, E. G., Schiff, J. E., Wilt, G. E., Hart, J. E., & James, P. (2021). Associations between nature exposure and health: a review of the evidence. *International Journal of Environmental Research and Public Health, 18*(9), 4790.

Joye, Y., & Dewitte, S. (2018). Nature's broken path to restoration. A critical look at Attention Restoration Theory. *Journal of Environmental Psychology, 59*, 1–8.

Kähkönen, J. E., Lionetti, F., Castelli, L., & Pluess, M. (2024). Development and validation of the highly sensitive child interview for the assessment of environmental sensitivity in primary school children. *Personality and Individual Differences, 222*, Article 112592.

Kaplan, S. (1995). The restorative benefits of nature: Toward an integrative framework. *Journal of Environmental Psychology, 15*, 169–82. https://doi.org/10.1016/0272-4944(95)90001-2

Li, Z., Sturge-Apple, M. L., & Davies, P. T. (2021). Family context in association with the development of child sensory processing sensitivity. *Developmental Psychology, 57*(12), 2165.

Lionetti, F., Aron, A., Aron, E. N., Burns, G. L., Jagiellowicz, J., & Pluess, M. (2018). Dandelions, tulips and orchids: Evidence for the existence of low-sensitive, medium-sensitive and high-sensitive individuals. *Translational Psychiatry, 8*(1), 24.

Lionetti, F., Aron, E. N., Aron, A., Klein, D. N., & Pluess, M. (2019). Observer-rated environmental sensitivity moderates children's response to parenting quality in early childhood. *Developmental Psychology, 55*(11), 2389.

Lionetti, F., Klein, D. N., Pastore, M., Aron, E. N., Aron, A., & Pluess, M. (2022). The role of environmental sensitivity in the development of rumination and depressive symptoms in childhood: a longitudinal study. *European Child & Adolescent Psychiatry, 31*(11), 1815–1825.

Lionetti, F., Dumpfrey, R. S. C., Richetin, J., Fasolo, M., Nocentini, A., Penolazzi, B., & Preti, E. (2024). Is environmental sensitivity a unique trait? A multi-sample study on the association between sensitivity, personality, and psychological adjustment. *Personality and Individual Differences, 217*, Article 112463.

Liss, M., Timmel, L., Baxley, K., & Killingsworth, P. (2005). Sensory processing sensitivity and its relation to parental bonding, anxiety, and depression. *Personality and Individual Differences, 39*(8), 1429–1439.

Mayer, F. S., & Frantz, C. M. (2004). The connectedness to nature scale: A measure of individuals' feeling in community with nature. *Journal of Environmental Psychology, 24*(4), 503–515.

Mayer, F. S., Frantz, C. M., Bruehlman-Senecal, E., & Dolliver, K. (2009). Why is nature beneficial? The role of connectedness to nature. *Environment and Behavior, 41*(5), 607–643.

Martin, L., White, M. P., Hunt, A., Richardson, M., Pahl, S., & Burt, J. (2020). Nature contact, nature connectedness and associations with health, wellbeing and pro-environmental behaviours. *Journal of Environmental Psychology, 68*, Article 101389.

McCormick, R. (2017). Does access to green space impact the mental well-being of children: A systematic review. *Journal of Pediatric Nursing, 37*, 3–7.

Moscardino, U., Scrimin, S., Lionetti, F., & Pluess, M. (2021). Environmental sensitivity and cardiac vagal tone as moderators of the relationship between family support and well-being in low SES children: An exploratory study. *Journal of Social and Personal Relationships, 38*(9), 2772–2791.

Nocentini, A., Menesini, E., & Pluess, M. (2018). The personality trait of environmental sensitivity predicts children's positive response to school-based antibullying intervention. *Clinical Psychological Science, 6*(6), 848–859.

Obradović, J., & Boyce, W. T. (2009). Individual differences in behavioral, physiological, and genetic sensitivities to contexts: Implications for development and adaptation. *Developmental Neuroscience, 31*(4), 300–308.

Ohly, H., Gentry, S., Wigglesworth, R., Bethel, A., Lovell, R., & Garside, R. (2016). A systematic review of the health and well-being impacts of school gardening: synthesis of quantitative and qualitative evidence. *BMC Public Health, 16*, 1–36.

Pluess, M. (2015). Individual differences in environmental sensitivity. *Child Development Perspectives, 9*(3), 138–143.

Pluess, M., & Belsky, J. (2013). Vantage sensitivity: individual differences in response to positive experiences. *Psychological Bulletin, 139*(4), 901.

Pluess, M., & Boniwell, I. (2015). Sensory-processing sensitivity predicts treatment response to a school-based depression prevention program: Evidence of vantage sensitivity. *Personality and Individual Differences, 82*, 40–45.

Pluess, M., Assary, E., Lionetti, F., Lester, K. J., Krapohl, E., Aron, E. N., & Aron, A. (2018). Environmental sensitivity in children: Development of the Highly Sensitive Child Scale and identification of sensitivity groups. *Developmental Psychology, 54*(1), 51.

Pluess, M., Lionetti, F., Aron, E. N., & Aron, A. (2023). People differ in their sensitivity to the environment: An integrated theory, measurement and empirical evidence. *Journal of Research in Personality, 104*, Article 104377.

Schaefer, M., Kevekordes, M. C., Sommer, H., & Gärtner, M. (2022). Of orchids and dandelions: Empathy but not sensory processing sensitivity is associated with tactile discrimination abilities. *Brain Sciences, 12*(5), 641.

Setti, A., Lionetti, F., Kagari, R. L., Motherway, L., & Pluess, M. (2022). The temperament trait of environmental sensitivity is associated with connectedness to nature and affinity to animals. *Heliyon, 8(7*), e09861.

Slagt, M., Dubas, J. S., van Aken, M. A., Ellis, B. J., & Deković, M. (2018). Sensory processing sensitivity as a marker of differential susceptibility to parenting. *Developmental Psychology, 54*(3), 543.

Song, C., Ikei, H., & Miyazaki, Y. (2016). Physiological effects of nature therapy: A review of the research in Japan. *International Journal of Environmental Research and Public Health, 13*(8), 781.

Sperati, A., Spinelli, M., Fasolo, M., Pastore, M., Pluess, M., & Lionetti, F. (2022). Investigating sensitivity through the lens of parents: validation of the parent-report version of the highly sensitive child scale. *Development and Psychopathology, 36*(1), 415–428.

Sperati, A., Acevedo, B. P., Dellagiulia, A., Fasolo, M., Spinelli, M., D'Urso, G., & Lionetti, F. (2024). The contribution of Sensory Processing Sensitivity and internalized attachment representations on emotion regulation competencies in school-age children. *Frontiers in Psychology, 15*, 1357808.

Sumner, R. C., Cassarino, M., Dockray, S., Setti, A., & Crone, D. M. (2022). Moving towards a multidimensional dynamic approach to nature and health: A bioavailability perspective. *People and Nature, 4*(1), 44–52.

Tam, K. P. (2013). Concepts and measures related to connection to nature: Similarities and differences. *Journal of Environmental Psychology, 34*, 64–78.

Ulrich, R. S. (1984). View through a window may influence recovery from surgery. *Science, 224*. https://doi.org/10.1126/science.6143402

Watts, N., Amann, M., Arnell, N., Ayeb-Karlsson, S., Beagley, J., Belesova, K., & Costello, A. (2021). The 2020 report of the Lancet countdown on health and climate change: responding to converging crises. *The Lancet, 397*(10269), 129–170.

Weyn, S., Van Leeuwen, K., Pluess, M., Lionetti, F., Greven, C. U., Goossens, L., & Bijttebier, P. (2021). Psychometric properties of the Highly Sensitive Child scale across developmental stage, gender, and country. *Current Psychology, 40*, 3309–3325.

Whitburn, J., Linklater, W., & Abrahamse, W. (2020). Meta-analysis of human connection to nature and proenvironmental behavior. *Conservation Biology, 34*(1), 180–193.

Wolf, M., Van Doorn, G. S., & Weissing, F. J. (2008). Evolutionary emergence of responsive and unresponsive personalities. *Proceedings of the National Academy of Sciences, 105*(41), 15825–15830.

World Health Organization. (2018). *Climate change and health.* World Health Organization. https://www.who.int/news-room/fact-sheets/detail/climate-change-and-health

Wilson, E. O. (1984). *Biophilia.* Cambridge: Harvard University Press.

Part III

The Resilience Strategies of Indigenous Communities

9

Indigenous Climate Inequalities: A Case Study with the Arctic Sámi Peoples

Reetta Toivanen

Setting the Scene: Sámi Indigenous Responses on the Growing Demand for Critical Minerals

> They say that they have their mines for the sake of the climate. I call bullshit! They have their mines to make money. Nobody opens a mine to save the climate. (Carter 2024).

This research was made possible by the Research Council of Finland funded Centre of Excellence in Law, Identity and the European Narratives decision number 1353311 and the Academy of Finland-funded project On 'Glocal' governance: On the meanings and consequences of the 'vernacularization' of Human Rights Concepts [grant number 256143], without any involvement in the content.

R. Toivanen (✉)
University of Helsinki, Helsinki, Finland
e-mail: reetta.toivanen@helsinki.fi

T. Skillington and A. Setti (eds.), *Climate Change Resilience Across Societal Contexts*, Energy, Climate and the Environment,
https://doi.org/10.1007/978-3-031-95020-9_9

This blunt statement by a Sámi interviewed by the journalist Bryan Carter in March 2024 encapsulates the contentious nature of mining, an industry often characterized by high stakes and dramatic shifts. As Theresa Komu (2019, 113) aptly observes, "Few industries are surrounded with such drama as mining: the excitement of a find, the devastation following a failure, the high risks and high hopes."

In March 2024, the European Union adopted a new law popularly called the Critical Raw Materials Act. The purpose of the act is, following Europe's commitment to green energy transition, to make sure that Europe is not only depending on critical minerals extracts in other parts of the world but in times of polycrises, EU has its own minerals secured. Many mining companies operating in Finland and Sweden were glad. This act means that much more mining activity will in future take place in the Northernmost parts of Europe. Also, in Sápmi, on the Sámi Indigenous peoples' homelands.

When the European Union passed the directive on Critical Raw Materials, it underscored the importance of ethical practices and community engagement. The directive mandates that projects must ensure engagement in good faith and comprehensive consultations with local communities, including Indigenous peoples (European Parliament, 2024, para 10). Annex III of the directive incorporates the principles of Free, Prior, and Informed Consent (FPIC) as established in the United Nations Declaration on the Rights of Indigenous Peoples, adopted by the UN General Assembly in 2007.

Further, paragraph 17 of the Critical Raw Materials Act stipulates that projects should ensure equitable consultations with relevant stakeholders, such as local communities and Indigenous peoples, with special attention to human rights, particularly in cases involving potential resettlement (European Union, 2024, 17). Paragraph 20 goes on to require that applications for projects potentially affecting Indigenous peoples must include a plan for meaningful consultations, minimization of adverse impacts, and fair compensation where appropriate (European Union, 2024, 20).

Despite these stipulations, the Factsheet on the law does not mention Indigenous peoples at all (European Union, 2023). In a recent OpEd,[1] Kaisa Rautio and Rasmus Klöcker Larsen (2023) argue that the Critical Raw Materials Act assumes Member State bureaucracies will consult in good faith with the Sámi. However, this assumption overlooks the deficiencies in national legislation in Sweden and Finland (and Norway, although it is not part of the EU), which lack effective provisions on Sámi rights in mineral laws and environmental licensing procedures (Rautio & Larsen 2023). This gap highlights the ongoing challenges in balancing economic interests with the rights and voices of Indigenous communities.

This article examines the conflict between the interests of Indigenous Sámi reindeer herders and various climate change mitigation measures. I will outline my methodology and introduce key theoretical frameworks, followed by a brief overview of the Arctic Sámi people. Subsequently, the global context of the Sustainable Development Goals (SDGs) will be discussed. This article aims at making a contribution to climate resilience by illustrating how Indigenous Sámi communities' rights, knowledge, and land-based practices are integral to mitigating and adapting to environmental change. By highlighting "green colonialism" and the marginalization of Indigenous perspectives in global sustainability frameworks, it wants to demonstrate that true climate resilience demands inclusive policies that recognize Indigenous stewardship and self-determination.

It is clear that land is a central issue for Indigenous peoples worldwide, and thus the Sámi people's relationship with land will be examined, before progressing to an analysis of competing land use from the perspectives of the Sámi and other local communities. The article concludes by asserting that Indigenous peoples manage substantial portions of the world's biodiversity-rich areas, and their stewardship is essential in preventing deforestation and protecting ecosystems. However, the SDGs fail to explicitly recognize the critical role Indigenous peoples play in biodiversity protection and climate change mitigation.

---

[1] An OpEd is a short opinion piece written by an external contributor of a journal that offers personal insights and perspectives on current issues.

Methods, Materials, and Ethical Approval

Methodology

The primary data for this article is drawn from ethnographic fieldwork conducted in the reindeer herding areas of Northern Finland, specifically in the region of Sápmi in Lapland, during the years 2013–2016.[2] This research involved in-depth interviews and participant observations, conducted either individually or collaboratively with colleagues, among Sámi reindeer herders and other key stakeholders within the community. The ethnographic approach was chosen to gain a nuanced understanding of the socio-ecological dynamics and land-use conflicts impacting the Sámi population. The interview and observation data were coded and analysed using Atlas.ti with a critical content analysis (Utt & Short, 2018).

Ethical Considerations and Positionality.

All interviews and observational data were collected in accordance with the United Nations principles of Free, Prior, and Informed Consent (FPIC), ensuring that participants were fully aware of the research objectives and voluntarily agreed to contribute. Moreover, the research strictly adhered to the ethical guidelines outlined by the Finnish National Board on Research Integrity (TENK, 2019), as well as the Ethical Guidelines of Sámi Research, approved in 2024 (Heikkilä et al., 2024).

As a Finnish researcher who grew up in the capital city, my positionality is important to acknowledge. I do not belong to the Sámi community, and my initial interest in Sámi issues emerged during my university studies when I sought to explore Sámi cultural rights from a human rights perspective. This positionality has influenced both the direction of the research and the interpretive lens through which the data is analysed.

[2] Academy of Finland-funded project On "Glocal" governance: On the meanings and consequences of the "vernacularization" of Human Rights Concepts.

Theoretical Framework and Key Concepts

Theoretically, this chapter is written between the disciplines of legal anthropology and Indigenous studies. In order to investigate the Indigenous responses to clean energy projects, the theoretical framework profits from the following key concepts: green colonialism, critical sustainability science, survivance, and resilience.

Green colonialism refers to the range of existing and emerging frameworks for climate change adaptation and mitigation that continue to perpetuate human rights violations and offer false solutions (Normann, 2020). These frameworks encompass, but are not limited to, initiatives such as "green" or "clean" energy, the development of a "green economy," energy transitions, and just transition strategies. Specific examples include projects like carbon capture technologies, industrial wind farms, solar power installations, mega-dams, nuclear energy, the introduction of invasive species, and geothermal developments (Zografos & Robbins, 2020). Additionally, practices such as greenwashing emissions reduction through carbon injection and the creation of "protected areas" are also included in this critique.

Critical Sustainability Science is an emerging interdisciplinary field that critiques and challenges mainstream approaches to sustainability, focusing on the power dynamics, inequities, and social injustices often embedded in global sustainability efforts (Krieg & Toivanen, 2021). It seeks to understand how sustainability practices, policies, and frameworks can perpetuate existing inequalities, marginalize vulnerable communities, or fail to address systemic issues, such as colonial legacies and capitalist exploitation, that undermine true sustainability.

Survivance, in this context, refers to the ontologies inherently connected to how Indigenous peoples have historically engaged with the world and their intrinsic rights to identity and existence (Castagno & Brayboy 2008; Tuck & McKenzie 2015; Vizenor 2008). Tuck & McKenzie (2015, 129), describe survivance as distinct from mere survival, characterizing it as "moving beyond basic survival in the face of overwhelming cultural genocide to create spaces of synthesis and renewal." Accordingly, survivance encompasses not only survival but also resistance. It is neither heroic nor tragic, but rather a means of overcoming both dominance and victimhood.

Moreover, the notion of survivance provides a more accurate lens through which to view the social ontologies that shape Indigenous peoples' lived experiences of climate change and land appropriation, as it foregrounds their agency, cultural continuity, and inherent rights to self-determination. In doing so, it highlights the limitations of mainstream resilience discourses, which often neglect the situated perspectives and experiences of local communities. By placing survivance at the centre, we can better recognize the complexities and particularities of Indigenous realities, and move towards a more inclusive and context-specific understanding of resilience.

The Arctic and the Sámi Indigenous Peoples

The Arctic is the northern polar region that encompasses Greenland, Iceland, and the northernmost areas of Russia, Canada, Alaska, and Scandinavia, including Svalbard. This region is characterized by unique climatic and ecological conditions, marked by extreme seasonal variations between summer and winter (Rantanen et al., 2022). Snow, ice, and permafrost are key elements that shape both hydrological and ecological processes, as well as influence the lives of those residing in the Arctic (Mustonen & Lehtinen, 2013). Approximately 4 million people inhabit the Arctic, around 10% of whom are Indigenous. Human presence in the Arctic dates back some 45,000 years, with over 40 distinct ethnic groups now living in the region, exhibiting significant social and cultural diversity. These populations have long adapted to the region's harsh conditions, developing specific social structures, technologies, and traditional knowledge integral to their livelihoods (Nysten-Haarala et al., 2021).

The ca 150 000 Sámi people live in four countries: Finland, Norway, Russia, and Sweden. In the aftermath of the Second World War, Sámi lands, Sápmi, became a focal point for the states, driven by geopolitical security concerns as well as the pursuit of economic interests, particularly in minerals and oil. This led to an increase in industrial activities and the exploitation of the region's resources, often described as a "resource frontier" (Toivanen & Fabritius, 2020).

In recent years, the effects of climate change have intensified environmental transformations in the Arctic, placing Indigenous, also the Sámi, communities at further risk. Changes in sea ice cover, extreme weather events, and ecological shifts, such as alterations in species distribution, have directly impacted their traditional livelihoods, which are closely tied to the utilization of terrestrial, marine, and freshwater resources, and are integral to both their culture and economies.

The Problem: Indigenous Peoples in the Margins of SDGs

It is widely acknowledged that human rights are fundamental to achieving sustainable development. This understanding was reinforced by the document Transforming Our World: The 2030 Agenda for Sustainable Development, adopted by the United Nations General Assembly on 21 October 2015 (A/RES/70/1). The core message of the Agenda is that all policies and processes aimed at implementing the Sustainable Development Goals (SDGs) should be grounded in human rights principles. Agenda 2030 reaffirms several key human rights commitments. In the section entitled "Our Shared Principles and Commitments," paragraph 10 states:

> The new Agenda is guided by the purposes and principles of the Charter of the United Nations, including full respect for international law. It is grounded in the Universal Declaration of Human Rights, international human rights treaties, the Millennium Declaration and the 2005 World Summit Outcome. It is informed by other instruments such as the Declaration on the Right to Development. (UN Agenda, 2030, para. 10)

The United Nations' Sustainable Development Goals (SDGs) consist of 17 goals designed to address various global challenges such as poverty, inequality, environmental degradation, and more, aiming to create a sustainable future by 2030 (ibid.). However, despite Indigenous peoples playing a crucial role in maintaining biodiversity, ecosystems, and local sustainable practices, their rights and knowledge systems are not sufficiently reflected in the SDGs (Toivanen and Cambou, 2021). They are only briefly acknowledged in two targets out of the 169 that comprise the SDGs:

Goal 2 (Zero Hunger): Target 2.3 addresses the need to double agricultural productivity and incomes of small-scale food producers, including Indigenous peoples. This is an important acknowledgement; however, it primarily frames Indigenous peoples in economic terms without deeper recognition of their cultural and environmental contributions.

Goal 4 (Quality Education): Target 4.5 emphasizes eliminating gender disparities in education and ensuring equal access to education. The target doesn't explicitly promote the protection or transmission of Indigenous knowledge systems, which are critical for sustainability and cultural survival.

The marginalization of Indigenous voices and knowledge in the global sustainability agenda, particularly within the framework of the Sustainable Development Goals (SDGs), raises concerns about potential consequences for both environmental and cultural outcomes (Toivanen & Cambou, 2021). Numerous studies have demonstrated that Indigenous communities often play a vital role in environmental stewardship, particularly in the management of ecosystems and biodiversity (see i.e., Heinämäki et al., 2014). Their traditional knowledge systems, developed over centuries of interaction with their environments, can offer valuable insights and approaches to contemporary sustainability challenges (Casi et al., 2021). These systems are often overlooked or underutilized in modern, technology-driven sustainability initiatives.

The limited inclusion of Indigenous rights and knowledges within the SDGs, with only two targets referencing Indigenous peoples, may constrain the global sustainability agenda's ability to fully address environmental and social issues. The above shown targets that mention Indigenous peoples are largely peripheral and do not engage with the deeper contributions that Indigenous practices and worldviews can provide to the sustainability discourse.

A stronger integration of Indigenous perspectives might enrich current strategies and contribute to more resilient and effective outcomes. Rather than being treated solely as recipients of development assistance, Indigenous peoples have the potential to serve as key partners in the formulation and implementation of sustainability policies. One issue that affects the sustainability and wellbeing of Indigenous peoples in most parts of the globe is climate change and the climate catastrophe that we are facing. Just now in October 2024, over 100 Indigenous organizations met in Geneva and issued a powerful statement titled "Just Transition: Indigenous Peoples' Perspectives, Knowledge, and Lived Experiences."[3] The statement sought to draw attention to the pernicious effects of climate change, as well as the adverse impacts of mitigation measures, on Indigenous communities. Studies show that insufficient attention has been paid to the ways in which climate change is expanding social inequalities between different societal cohorts and increasing vulnerability to experiences of harm (Lancet, 2022). In the case of Arctic peoples, such as Sámi, climate change impacts are occurring three to four times faster when compared with the rest of the world (Näkkäläjärvi et al., 2023). In the following subchapter, the significance of land for Indigenous survivance is examined, with the Sámi people serving as a case study.

What Matters is Land

As Christina Allard (2011) has described in rich detail, land rights form the very essence of Indigenous world-views and therefore lie at the centre of their demands for rights, including among the Sámi. Regarding ownership, states implicitly or explicitly lay claim to Sápmi, the expansive transnational territory constituting the customary homeland of the Sámi people.

When I asked people in Inari about the kind of tensions that are there between groups of people and what kinds of issues could be sources of conflict in Inari, the answer was thorough:

[3] https://www.indigenoussummit.org/summit-outcome (accessed in October 2024).

> Well, land use issues are one thing, meaning the ILO [196] Convention is sort of about cultural self-governance, so yes, that shows. What I see is that it's jealousy, it's nothing else, and the other thing is the ignorance that comes with it. Reindeer herding is, of course, a very significant livelihood here, and then, [...], if you think about what could be related to it, then in the Lemmenjoki area, mining or gold panning is one thing that is a problem. Logging is a problem in the sense that it also narrows the grazing areas, and then if you're personally an employee or an official, it's quite natural that salaried employees negotiate and stand up for their rights and want to promote their own interests. But if a reindeer herder or a practitioner of a traditional livelihood stands up for their own interests, then it's like something completely astonishing to the majority, that they defend their own livelihood. So yes, you asked about what is also a cause of conflict, or I don't know if it's a cause of conflict, but a threat—I actually talk about these as threats— some mining activities, logging, land use [...]. (Glocal_inari_16female_6.2013)

Although the answer to the question above is only one example, it illustrates how most local residents, especially those whose livelihoods and culture depend on reindeer herding, think. Reindeer herding is threatened by several competing land uses, and the profits from these other activities largely go to people who live outside Indigenous communities.

In Finland, not only the Sámi home area, but most of the forests (55%) in Lapland are owned by the state and managed by the state own forest company called Metsähallitus (Korhonen et al., 2021). Metsähallitus has practiced a ruthless forest policy since the Second World War: after the war, Finland had to pay war compensation to Soviet Union. In that endeavour, one third of the war reparations (approx. 300 million dollars) was paid in timber and forestry products (Turunen et al., 2018). The wood was also taken from Lapland even though due to the harsh climate; trees grow there much slower. Taking timber from Lapland was as if taking it from no man's land because Finland considered Lapland as a noninhibited area, a periphery without relevant population (Toivanen, 2019).

Land use and competing land use is the largest source of problems in the Arctic area. At the same time, according to the SGDs but also the whole climate movement, the search for alternative sources for energy, the green and clean deal alike, all need land. These developments impact the Sámi and their reindeer herding. They therefore face pressures from different sites, including not only climate change itself but also mitigation strategies, such as the development of renewable energy (Cambou, 2020; Normann, 2020).

Studies on extractivism underscore that extractive activities cannot be conducted sustainably, as they rely on non-reciprocal relationships between humans and non-human entities, leading to environmental exploitation (Kröger et al., 2021). Sámi worldviews, however, are deeply rooted in a profound connection to their lands, nature, and a respect for the balance among all living beings (Helander, 2000). Despite this, contemporary discourses often frame this relationship through a dichotomous separation between nature and culture, a conceptual divide that has not historically existed in many Sámi worldviews (Guttorm, 2021, 237).

An Indigenous perspective is founded on a mutual relationship between humans and other living entities (Jääskeläinen, 2020, 50). This worldview stands in stark contrast to the fundamental issue in the approval of mining projects, which are typically executed based on the perspectives of developers or governmental authorities. These processes operate under the assumption that Sámi communities will adapt to industrialization, inherently privileging non-Indigenous ways of defining significant impacts (Jääskeläinen, 2020, 52).

State and corporate stakeholders have exploited this discourse to marginalize Indigenous peoples, positioning them outside the realm of culture and, consequently, outside the sphere of political agency (Toivanen, 2019). Concurrently, national majority literature and state stakeholders have frequently portrayed the Arctic region as peripheral and devoid of inhabitants, while simultaneously depicting it as a treasure trove ripe for resource extraction (Eriksen, 2018). These narratives continue to marginalize and silence the local Indigenous peoples of the Arctic today.

And if this land is at the same time the heart of cultures and languages of a minority people, we face a problem: is the continuation of Indigenous peoples less important than clean energy?

Competing Land Use: Clean Energy through Wind Parks and Mines

In its Arctic strategy, the European Commission (2021, 9) acknowledges Arctic states as key suppliers of raw materials and asserts that the European Green Deal (EGD) is pivotal to the EU's Arctic engagement (ibid, 1). Concurrently, climate change is progressing more rapidly in the Arctic than in other regions (Sörlin, 2022, 14), and geopolitical shifts are increasing interest in the Arctic and its resources (ibid, 18).

In addition to mining for critical minerals, wind energy has been presented as one possible solution in order to get carbon-free energy. However, wind parks have their problems, too: the Sámi reindeer herders and hunters point out that wind turbines have been found to disturb larger mammals, such as reindeers, leading to habitat loss and decreases in population size (Helldin et al., 2012; Skarin & Åhman, 2014). For example, Sweden aims to achieve a 100% renewable energy production by 2045, which requires an expansion in technologies such as wind power (Cambou, 2020). The amount of wind power in Sweden has therefore increased strongly, with an exponential growth in the northernmost counties (ibid.).

The Sámi Parliament in Finland views the Agenda 2030 and the Sustainable Development Goals (SDGs) with a critical lens, particularly regarding their lack of recognition for the rights and needs of Indigenous peoples. While the goals of sustainable development align with broader societal objectives, the Sámi Parliament has expressed concerns that the SDGs fail to sufficiently account for the unique position of the Sámi people, Finland's only Indigenous group. One key criticism is that the Agenda 2030 framework tends to prioritize state-driven policies and solutions without adequately consulting or involving Indigenous communities, which can undermine their self-determination and cultural preservation. The Sámi Parliament in the same manner as the UN Permanent Forum for Indigenous Issues emphasizes that sustainable development for Indigenous peoples must go beyond environmental and economic considerations to include the protection of their languages, traditions, land rights, and governance structures (Toivanen & Cambou, 2021). The Parliament has also highlighted that the implementation of SDGs in Finland often overlooks the specific socio-cultural and economic challenges faced by the Sámi, calling for a more inclusive approach that respects Indigenous rights and international commitments such as the UN Declaration on the Rights of Indigenous Peoples (UNDRIP) (Näkkäläjärvi et al., 2018).

In general, the international organizations such as UN and ILO do not enjoy trust in the Finnish Lapland. People seem to be rather tired with human rights activists coming from the South and trying to "help." They are considered to lack understanding of the complexity of Sámi people and their cultures. As one Sámi woman said:

> But I don't really know, in my opinion, human rights organizations and such aren't, this is like, actually, I would say that if there's some big organization whose active members are still mostly people from outside the North or, of course, also Sámi people, then they probably have a very stereotypical view of Sámi culture and who the Sámi are, what they do, and what their culture includes. So, I actually think that they should almost stay completely out of those issues because they don't understand them. (Glocal_inari_15female_6.2013)

I talked with a group of reindeer herders about their possibilities to influence decision-making in land-use matters. Thus, I asked who according to them holds the power in matters concerning Sámi affairs and issues related to Sámi reindeer herders?

One responded:

> The power in matters concerning Sámi reindeer herding certainly does not lie with the Sámi; it lies elsewhere. When it comes to cultural matters, the Sámi Parliament has authority over culture, language, and other related issues. However, in terms of the reindeer herding industry and Sámi reindeer herding culture, the power lies with others due to the structure of the Reindeer Herders' Association. (Glocal_inari_17male_6.2013)

I continued asking about his impression on his own impact in decision-making and contacts with decision-makers. He replied:

> Yes, there are connections. We have been in contact with them and have collaborated, and some things have progressed and improved. However, it feels like some matters never move forward, giving the impression that we are heard but not listened to. (Glocal_inari_17male_6.2013)

When asked, what should happen in order to increase the voice of reindeer herders and other people in natural substance economies in Lapland. The answer came without a pause from a reindeer herding woman:

> That for once, everyone would sit at the same table and see who actually has the rights. And not just talk about how Indigenous peoples have this or that right. (Glocal_inari_20male_6.2013)

When I asked my interlocutor whether the competing land use, building windmill parks, or mines threaten the natural substance economies already, she answered:

> Yes, it is threatened because reindeer herding is the cornerstone of Sámi culture. It carries knowledge and language passed down through generations, encompassing valuable traditional knowledge. Reindeer have multiple dimensions, such as food, handicrafts, and nature. Therefore, I would say that reindeer herders have a fundamentally different relationship with nature compared to, for instance, a Sámi teacher. Moreover, I find it deeply concerning that our young people do not know how to navigate the forest. They cannot read the signs of nature, catch fish, or know the paths reindeer take. Reindeer herding is our final link to nature and its sustainable utilisation. (Glocal_inari_8female_6.2013)

For people living in the Sámi home area in Finland, the Finnish legal system still protects the area from extensive mining or wind park projects, but it will be a matter of time when the EU directive on critical raw materials discussed earlier will impact also this area. Meanwhile, Sámi and local communities residing just beyond the boundaries of the legally protected homeland have long experienced continuous resource extraction and exploitation, with far fewer legal safeguards to protect their rights and livelihoods.

Concluding Discussion

> It feels like 'live there if you want, but you won't be given anything.' It's kind of like 'you can stay there, but only if you manage on your own.' So, if Finland is still considered a welfare state, it doesn't extend everywhere. (Glocal_inari_12female_6.2013)

In this article, the main aim was to show how insufficiently Indigenous rights are recognized in the discussions about sustainability and climate change: The SDGs do not comprehensively reflect Indigenous peoples' rights, which are outlined in the UN Declaration on the Rights of Indigenous Peoples (UNDRIP). Indigenous groups have long fought for recognition of their land rights, sovereignty, and self-determination, yet these rights are marginalized in global frameworks like the SDGs.

The other issue is the lack of inclusion of traditional knowledge in the sustainability and climate debates, leading too easily to green colonialism. Green colonialism here means that the competing land use between traditional livelihoods, mining, and wind or solar energy projects insufficiently or not at all takes Indigenous knowledge into consideration. Indigenous knowledge and practices are often key to sustainable environmental management, including forest conservation, water management, and biodiversity protection. However, the SDGs largely focus on modern scientific or technological solutions to sustainability challenges, sidelining the vast contributions of Indigenous ecological knowledge.

The drive for a green transition, essential for global sustainability, must not come at the cost of Indigenous communities and local populations, who maintain longstanding, intergenerational, and diverse socio-ecological relationships with Arctic lands. Instead, decisions about new clean energy projects should be collaboratively developed with local stakeholders, taking into account their historical marginalization and the cumulative effects of resource extraction.

The discussion surrounding the effectiveness and limitations of clean energy projects in the Arctic is still in its infancy, highlighting the urgent need for further research to guide a just transition that respects the interests of all involved parties. As we work towards the targets set out in

Agenda 2030, it is crucial not to reinforce existing power disparities or marginalize vulnerable communities. By placing Indigenous perspectives at the centre and prioritizing social inclusion, we can create a fairer and more sustainable future for all who live in the Arctic region.

One of the key conclusions is that Indigenous peoples manage large portions of the world's biodiversity-rich territories. Their stewardship has been shown to be critical in preventing deforestation and protecting ecosystems. Yet, the SDGs fail to directly address the critical role Indigenous peoples play in the protection of biodiversity and in climate change mitigation. Thus, one question remains to be addressed: Shouldn't a sustainable development in the Arctic be steered by the same principles of environmental justice and human rights as elsewhere?

Acknowledgements This research was made possible by the Research Council of Finland funded Centre of Excellence in Law, Identity and the European Narratives decision number 1353311 and the Academy of Finland-funded project On "Glocal" governance: On the meanings and consequences of the "vernacularization" of Human Rights Concepts [grant number 256143], without any involvement in the content.

References

Allard, C. (2011). The Nordic countries' law on Sámi territorial rights. *Arctic Review on Law and Politics, 2*(2), 159–183.

Cambou, D. (2020). Uncovering injustices in the green transition: Sámi rights in the development of wind energy in Sweden. *Arctic Review on Law and Politics, 11*, 310–333. https://doi.org/10.23865/arctic.v11.2293

Casi, C., Guttorm, H. E., & Virtanen, P. K. (2021). Traditional ecological knowledge. In C. Krieg & R. Toivanen (Eds.), *Situating sustainability*. Helsinki University Press. https://doi.org/10.33134/HUP-14-13

Castagno, A. E., & Brayboy, B. M. J. (2008). Culturally responsive schooling for indigenous youth: A review of the literature. *Review of Educational Research, 78*(4), 941–993. https://doi.org/10.3102/0034654308323036

Carter, B. (2024, May 3). Sweden's Sámi pay the price in Europe's quest for raw materials. *EuroNews*. https://www.euronews.com/2024/05/03/swedens-sami-pay-the-price-in-europes-quest-for-raw-materials.

Eriksen, T. H. (2018). The paradox of autonomy. In H. T. Eriksen, S. Valkonen, & J. Valkonen (Eds.), *Knowing from the Indigenous North* (pp. 67–72). Routledge.

European Parliament (2024). P9_TA(2023)0325 Framework for ensuring a secure and sustainable supply of critical raw materials Amendments adopted by the European Parliament on 14 September 2023 on the proposal for a regulation of the European Parliament and of the Council establishing a framework for ensuring a secure and sustainable supply of critical raw materials and amending Regulations (EU) No 168/2013, (EU) 2018/858, (EU) 2018/1724 and (EU) 2019/1020 (COM(2023)0160 – C9–0061/2023 – 2023/0079(COD)).

Union, European. (2023). *Factsheet on the critical raw materials act*. Publications Office of the European Union.

European Commission. (2021). Joint Communication: A stronger EU engagement for a peaceful, sustainable and prosperous *Arctic*. JOIN (2021) Final 27. https://www.eeas.europa.eu/sites/default/files/2_en_act_part1_v7.pd

European Union (2024). Critical Raw Materials Act. Regulation (EU) 2024/1252 of the European Parliament and of the Council of 11 April 2024 establishing a framework for ensuring a secure and sustainable supply of critical raw materials and amending Regulations (EU) No 168/2013, (EU) 2018/858, (EU) 2018/1724 and (EU) 2019/1020.

Guttorm, H. (2021). Becoming earth: Rethinking and (Re-)Connecting with the earth, Sámi lands and relations. In R.-H. Andersson, B. Cothran, & S. Kekki (Eds.), *Bridging cultural concepts of nature: Indigenous people and protected spaces of nature* (pp. 229–258). Helsinki University Press.

Heikkilä, L., Kuokkanen, R., Lehtola, V-P., Magga, P., Magga, S.-M., Näkkäläjärvi, J., Valkonen, S., & Virtanen, P. (2024). *Ethical guidelines for research involving Sámi people in Finland. Giellagas-instituutti.* University of Oulu. https://rightsstatements.org/vocab/InC/1.0/

Heinämäki, L., Herrmann, T. M., & Neumann, A. (2014). The protection of the culturally and spiritually important landscapes of arctic indigenous peoples under the convention on biological diversity and first experiences from the application of the Akwé: Kon guidelines in Finland. *The Yearbook of Polar Law Online, 6*(1), 189–225. https://doi.org/10.1163/1876-8814_008

Helander, E. (2000). Saamelainen maailmankuva ja luontosuhde. Beaivvi mánát Saamelaisten juuret ja nykyaika. *Tietolipas, 164*.

Helldin, J.-O., Jung, J., Neumann, W., Olsson, M., Skarin, A., Widemo, F., Sverige, Naturvårdsverket, & Vindval (Program). (2012). *The impacts of wind power on terrestrial mammals: A synthesis*. Naturvårdsverket.

Jääskeläinen, T. (2020). The Sámi reindeer herders' conceptualizations of sustainability in the permitting of mineral extraction–contradictions related to sustainability criteria. *Current Opinion in Environmental Sustainability, 43*, 49–57.

Komu, T. (2019). Dreams of treasures and dreams of wilderness–engaging with the beyond-the-rational in extractive industries in northern Fennoscandia. *The Polar Journal, 9*(1), 113–132.

Korhonen K. T., Ahola A., Heikkinen J., Henttonen H. M., Hotanen J.-P., Ihalainen A., Melin M., Pitkänen J., Räty M., Sirviö M., Strandström M. (2021). Forests of Finland 2014–2018 and their development 1921–2018. *Silva Fennica, 55* (5) article id 10662. https://doi.org/10.14214/sf.10662

Krieg, C., & Toivanen, R. (Eds.). (2021). *Situating sustainability*. Helsinki University Press. https://doi.org/10.33134/HUP-14-13.

Kröger, M., Hagolani-Albov, S., & Gills, B. (2021). Extractivisms. In C. P. Krieg & R. Toivanen (Eds.), *Situating sustainability: A handbook of contexts and concepts* (pp. 239–252). Helsinki University Press.

The Lancet Report (2022). The 2022 report of the Lancet Countdown on health and climate change: health at the mercy of fossil fuels. Romanello, Marina et al. *The Lancet, 400*(10363), 1619–1654.

Mustonen, T., & Lehtinen, A. (2013). Arctic earthviews: Cyclic passing of knowledge among the indigenous communities of the Eurasian North. *Sibirica: Journal of Siberian Studies, 12*(1). https://doi.org/10.3167/sib.2013.120102.

Näkkäläjärvi, H. K., Jaakkola, J., & Juntunen, S. (2018). The holistic effects of climate change on the culture, well-being, and health of the Saami, the only indigenous people in the European Union. *Current Environmental Health Reports, 5*(4), 401–417. Article 211. https://doi.org/10.1007/s40572-018-0211-2

Näkkäläjärvi, K., Juntunen, S., & Jaakkola, J. J. K. (2023). Ilmastopolitiikan oikeudenmukaisuus alkuperäiskansa saamelaisten kannalta ['Finnish climate panel, justice in climate policy from the viewpoint of Indigenous Sami people']. *Suomen ilmastopaneelin julkaisuja*.

Normann, S. (2020). Green colonialism in the Nordic context: Exploring Southern Saami representations of wind energy development. *J Community Psychol., 49*, 77–94. https://doi.org/10.1002/jcop.22422

Nysten-Haarala, S., Joona, T., & Hovila, I. (2021). Wind energy projects and reindeer herders' rights in Finnish Lapland: A legal framework. *Elementa, 9*(1). https://doi.org/10.1525/elementa.2020.00037

Rantanen, M., Karpechko, A. Y., Lipponen, A., Nordling, K., Hyvärinen, O., Ruosteenoja, K., & Laaksonen, A. (2022). The Arctic has warmed nearly four times faster than the globe since 1979. *Communications Earth & Environment, 3*(1), 1–10.

Rautio, K., & Larsen Kloecker, R. (2023, November 10). *EU's critical raw materials act fails to protect Sámi rights—Here's how to strengthen it*. EURACTIV. https://www.euractiv.com/section/energy-environment/opinion/eus-critical-raw-materials-act-fails-to-protect-sami-rights-heres-how-to-strengthen-it/. Accessed in October 2024.

Skarin, A., & Åhman, B. (2014). Do human activity and infrastructure disturb domesticated reindeer? The need for the reindeer's perspective. *Polar Biology, 37*(7), 1041–1054. https://doi.org/10.1007/s00300-014-1499-5

Sörlin, S. (2022). The extractivist paradigm: Arctic resources and the planetary mine. In S. Sörlin (Ed.), *Resource extraction and arctic communities: The new extractivist paradigm* (pp. 3–32). Cambridge University Press.

TENK (2019). *Guidelines for ethical review in human sciences*. Finnish National Board on Research Integrity 3/2019 (2nd, revised edn). https://tenk.fi/sites/default/files/2021-01/Ethical_review_in_human_sciences_2020.pdf

Toivanen, R. (2019). European fantasy of the arctic region and the rise of indigenous Sámi voices in the global arena. In N. Sellheim, Y. Zaika, & I. Kelman (Eds.), *Arctic Triumph. Springer Polar Sciences*. Springer. https://doi.org/10.1007/978-3-030-05523-3_3

Toivanen, R., & Cambou, D. (2021). Human rights. In C. Krieg & R. Toivanen (Eds.), Situating sustainability (pp. 51–62). Helsinki University Press. https://doi.org/10.33134/HUP-14-13.

Toivanen, R., & Fabritius, N. (2020). Arctic youth transcending notions of 'culture' and 'nature': Emancipative discourses of place for cultural sustainability. *Current Opinion in Environmental Sustainability, 43*, 58–64. https://doi.org/10.1016/j.cosust.2020.02.003

Tuck, E., & McKenzie, M. (2015). Indigenous methods and critical place inquiry. In M. McKinley (Ed.), *Place in research: Theory, methodology, and methods* (pp. 129–142). Routledge. https://doi.org/10.4324/9781315764849-6

Turunen, M. T., Rasmus, S., & Kietäväinen, A. (2018). The importance of Reindeer in Northern Finland during World War II (1939–45) and the Post-War Reconstruction. *Arctic, 71*(2), 167–182. https://doi.org/10.14430/arctic4717

Utt, J., & Short, K. G. (2018). Critical content analysis: A flexible method for thinking with theory. *Understanding and Dismantling Privilege, 8*(2), 1–7. Retrieved from https://wpcjournal.com/article/view/18826

Vizenor, G. (2008). Aesthetics of survivance. In G. Vizenor (Ed.), *Survivance: Narratives of native presence* (pp. 1–25). University of Nebraska Press.

Zografos, C. & Robbins, P. (2020, November 20). Green sacrifice zones, or why a green new deal cannot ignore the cost shifts of just transitions. *One Earth, 3*(5), 543–546. https://doi.org/10.1016/j.oneear.2020.10.012

10

Human-Reindeer Well-Being Under Threat in the Arctic: The Case of Kola Peninsula Indigenous Communities

Vladislava Vladimirova

Introduction

Reindeer husbandry has been singled out as a uniquely human adaptation in the Arctic, utilizing a resource-poor environment over huge areas in harsh climatic conditions. It is grounded in highly specialized cultural and environmental knowledge and skills. The limited success of 20th-century experiments at "modernizing reindeer economy," enforced by scientists and state planners to improve the efficiency of resource use, production, breeds, and methodology, has shown that Indigenous knowledge and relations to reindeer are the indispensable core (Beach, 1981). The survival of reindeer husbandry, irrespective of economic adjustments and planners' ambitions, is premised on these core relations and knowledge shifts that adjust and resist unwise modernization efforts and deteriorating environmental factors. Soviet planners conducted a

V. Vladimirova (✉)
Uppsala University, Uppsala, Sweden
e-mail: vladislava.vladimirova@ires.uu.se

T. Skillington and A. Setti (eds.), *Climate Change Resilience Across Societal Contexts*, Energy, Climate and the Environment,
https://doi.org/10.1007/978-3-031-95020-9_10

grandiose effort of administration and economic development of the Arctic, conducting mass-scale collectivization of reindeer and reforming the rationale, organization, methodology, and infrastructure of reindeer herding and biology, behavior, and conception of reindeer. In the process, Indigenous practitioners were converted into urban employees who studied herding at village schools. This transformation had diverse effects on Indigenous cultures.

In the postsoviet period, reindeer herding in the Russian Arctic attracted attention due to the opening of the region for global research, the severe problems or even collapse of some Indigenous reindeer enterprises, and the revival of others. Scholars looked for causes behind different local transformations and possible actions to strengthen and revive reindeer herding. This led to attention to the inherent economic, ecological, and cultural values of the reindeer/human nexus (Krupnik, 2000). Some saw the decrease in reindeer numbers as an ecological adaptation and a fix to Soviet-era production increases that had exceeded the carrying capacity of the range and defied traditional knowledge (Anderson, 2006).

Studies of reindeer husbandry in Russia take an anthropocentric perspective and focus either on the economic dimension or the religious and symbolic aspects of reindeer. Anthropologies of reindeer pastoralism are looking for metaphors that can better capture the multidimensional relations, like symbiotic domestication with humans (Beach & Stammler, 2006), living in a sentient ecology (Anderson, 2000), and symbiotic domesticity (Stammler, 2010). David Anderson describes similar metaphors of reindeer personhood as inclusive but "overwriting Northern voices" (Anderson, 2017, p. 134). Drawing on Indigenous languages and ecological descriptions of place, on the one hand, and on landscape ecologists' and archaeologists' conceptualizations of "landscapes of relationships…wherein people and animals come to know each other and respect each other's copresence," Anderson suggests the concept of domus, "an emplaced sense of being, or a familiar landscape … as a holistic ensemble, which guides action of Arctic humans and nonhuman persons" (Ibid., 142). In this frame, Indigenous communities' well-being depends on the thriving of this emplaced relationship of humans and reindeer living together or the well-being of the domus.

Scholars show that applying the conventional United Nations Human Development Index with three components: life expectancy, education, and gross domestic product per capita provides little understanding of Arctic Indigenous communities' welfare. While their index shows low scores, many individuals in this region exhibit a strong sense of well-being. To provide more nuanced measurement, scholars identified three additional domains that are crucial for Arctic residents: *fate control* refers to guiding one's destiny in a region where both climate and human mobility are in constant and unpredictable flux; *cultural vitality* refers to Indigenous people belonging to a viable local culture with shared beliefs, norms, and common history; and finally *contact with nature*, which refers to people's access to and interaction with the natural world (Larsen et al., 2014). These indicators constitute an essential minimum for Indigenous well-being and their assessment can help identify stressors and correlate with resilience (ibid.). In this study, however, following reindeer herders' wisdom and Anderson's notion of domus, I suggest a post-human perspective on well-being that incorporates the flourishing of the whole herding "ensemble," human and non-human persons and place (Anderson, 2017). This contrasts with the concept of resilience, that is often used to justify governing domus in ways that transform and displace it.

In this analysis, I employ post-structuralist criticism of resilience and a frame of biopolitics developed in the analysis of nature conservation and climate change interventions in political ecology. Political ecologists study interventions in the context of biodiversity preservation and climate change adaptation, where different stakeholders (academics, advisors, NGOs, businesses) are orchestrated by state authorities as part of biopolitics. Biopolitics is often used after its original definition in Foucault's works as techniques and rationalities that governments employ to shape individual and collective life in the name of alleged positive ends, such as development, safety, and prosperity (Foucault, 2003). Biopolitics includes forms of governance that work through correction, exclusion, disciplining, optimization, and normalization. The objects of biopolitical interventions are life itself, and their justifications are its enhancement (Grove, 2014). The governance of life itself, as in Foucault's famous quote, involves managerial decisions about "making live" or "letting die." Political ecologists have further expanded the scope of biopolitics from human populations to include "asymmetrically valued forms of both human and non-human lives within rapidly shifting ecological conditions," which are of special relevance to the analysis of reindeer herding domus (Cavanagh, 2014, p. 277).

In the following section, I discuss critically the Arctic resilience frame in relation to reindeer herding, and I sketch the evolving Russian debate on resilience. I then extend my analysis to governance of reindeer herding and Indigenous people in the Kola Peninsula, with a specific focus on one protected area, Seid"iavvr' zakaznik.[1] I use ethnography collected over repeated field trips between 2001 and 2017 and published sources and documents.

Arctic Resilience

Resilience as an extension of sustainable development is suggested as a "solution" to expanding negative impacts of climate change and industrialization on Arctic communities. Resilience is defined as the capacity of social-ecological systems to cope with disturbances by "responding or reorganizing in ways that maintain essential identity, function, and structures, as well as the capacity to navigate and shape change" (Carson & Peterson, 2016, p. xvi). Critiques point out that sustainable development, according to its proponents, extends the notion of development to include both economy and diverse understandings of human well-being, thus shifting focus from development of human life to non-human "life support systems" (Reid, 2013). Resilience is instrumental in shaping governmental practices, institutions, and rationalities, and produces subjectivities which envisage and constitute the world in terms of "disastrousness," and subjects as "called upon to bear the disaster" in order to "grow and prosper in the world" (ibid., 355). Resilience thus gives birth to political subjects intent upon changing the world. The resilient

[1] *Zakaznik* is a protected territory in Russia where certain economic activities are permitted to a limited extent; it corresponds to IUCN III, Natural Monument, or VI, Protected Area with sustainable use of natural resources.

subject also receives continuous responsibility to "adapt" and "change" in response to threats and shocks, i.e., to be the agent of one's own well-being against conditions that one has no agency to change (ibid.).

Arctic resilience theorists stress the role of learning as part of adapting to change and necessary system transformations (Carson & Peterson, 2016). Learning is a solution to "ecological ignorance," a deficiency attributed to "vulnerable" groups, and to climate change threats, not only to the biosphere but to all humanity. Poor and vulnerable groups are said to need to learn sustainable use of "ecosystem services" to become resilient subjects (Reid, 2013). The Arctic Resilience Report acknowledges the negative historical impact of colonial policies in the Arctic, where limited movement, forced schooling, bans or limits on economic activities, and unequal rights to land and resources have weakened Indigenous peoples' resilience. Both the Arctic region and Indigenous people are described as "vulnerable" and in need of resilience "enhancement." In Russia, loss of resilience is described in the case of Teriberka, Kola Peninsula, where Soviet planners replaced reindeer herding livelihoods with industrial fisheries. This is said to have reduced capacity to develop alternative strategies due to traditional knowledge loss, which is reflected in present-day poverty of the village and its residents (Carson & Peterson, 2016).

In the report, Arctic resilience is justified through a historical detour via the claim that this capacity has always been important for communities living in the Arctic and maintaining their social-economic systems in an unpredictable climate. The report showcases the high resilience of Yamal-Nenets, an emblematic tundra reindeer herding group, who are praised for preserving strong cultural traditions and intensifying reindeer husbandry, adapting and transforming to solve problems they encounter, such as climatic variability, postsoviet political and economic crises, and expanding industrial development (Carson & Peterson, 2016). This analysis fails to acknowledge that Soviet planners took traditional Nenets tundra reindeer husbandry as a model which they attempted to impose on other groups, whose traditional methods, environmental conditions, and reindeer breeds were incompatible with Nenets practices.

Even Nenets resilience, however, according to the report, does not suffice at the present moment, as climate and human activities linked to the global economy are accelerating in complex ways the transformation of the Arctic. Such processes are claimed to justify mainstreaming resilience in global Arctic policy under the auspices of the Arctic Council-initiated assessment process. The Arctic Resilience Report identifies external sources of risk, such as climate change, the long-distance transport of air pollutants, and long-distance movement of animal and plant populations and ecosystems. Reindeer herding communities, according to the report, express increasing concerns about the negative impacts of rising mineral resource extraction, industrial production, tourism, and geopolitical conflicts that encroach on grazing territories. Later works, perhaps in an effort to "correct" the colonial stereotypes implicit in the report, state that postcolonial resilience thinking needs local approaches, broader deliberation, and democratic participation for the co-production of knowledge with vulnerable and affected Indigenous communities in order to "arrive at new development pathways" (Nilsson & Larsen, 2020).

Resilience thinking is critiqued as a tool for justifying and maintaining neoliberal markets, systems of governance, and subjectivities against radical reappraisals and alternative economic and political orders. Sustainable development emphasizes the urgent need to secure biosphere life, and neoliberal politicians apply market mechanisms to ensure security. The biosphere itself is conceived not as threatened by further economic growth, but as an economy of "services" that "humanity receives." (Reid, 2013). Property rights regimes should be extended to incorporate ecosystem services and enable markets to function in them. Resilience is thus a powerful tool that legitimizes, naturalizes, and spreads neoliberal governance and institutions globally (Reid, 2013).

Under the influence of global institutions, like UNEP, of which it is a member, Russia has embraced the language and policies of resilience. Russian theorists both emulate global vocabulary and downplay their own thematic fields and concepts. Territorial resilience, the study of urban and regional planning and stability of governance, has received much attention (Nenasheva, 2023; Odintsova, 2020). Russian studies emphasize the ability of the system to (a) anticipate and prevent, (b) resist and absorb, (c) react, (d) adapt, and (e) re-cover. Paradoxically, Russian literature on resilience is less engaged with the Arctic, despite its largest part being in Russia. One study that does examine the resilience of Arctic socioeconomic systems in the context of global warming and increased extreme climatic phenomena recommends zoning territory according to levels of risk, as well as the development of regulatory documents governing the life of the population in the region (Vinogradova, 2021). Arctic communities are envisioned as lacking social integration due to the great ethnic and cultural diversity. The resilience of Arctic communities is described as highly "dependent on the local context and is not related to the resilience of other communities" (Nenasheva, 2023). Arctic communities' resilience has internal sources, such as local knowledge, culture, and social organization, that provide some strategies for adaptation. Nevertheless, community internal resources are not capable of providing strategies for dealing with all external influences, and external resources, primarily local and regional authorities, have greater power to assess threats and provide coordinated plans and finances for activities in the community. Russian Arctic resilience is thus centralized and requires orchestrated development of strategies to respond to threats (ibid.). Other studies add to this centralized approach to resilience through developing its main premises: diversifying the economy; improving institutional conditions for business; security; stable infrastructure; attention to environmental problems; and finally social capital that is able to understand, evaluate, and minimize effects of external shocks. Last but not least in this enumeration, resilience is not a backward movement toward a previous state but a process of transformation and adaptation (Odintsova, 2020).

Kola Reindeer Herding

On the Kola Peninsula, Soviet reindeer enterprises united two ethnic groups and herding traditions, Sami and Komi. Soviet planners drastically transformed the social, spatial, and temporal organization of herding, as well as Indigenous people's relations to it. Under new conditions, the labor of herding in the tundra continued to be performed by Indigenous practitioners or those who learned skills from them. Reindeer, in many cases, migrate seasonally over huge territories, and reindeer herders' work could not be monitored and guided to the same degree as other employees. Despite the drastic transformation of herding practice and rationale, its domus has been vital and played an important role for Indigenous communities and cultures in the Kola Peninsula. In the postsoviet period, however, herds rapidly decreased. Most of the losses were blamed on poaching—in different sources, the figures vary from 1000 to 5000–6000 lost animals per year. An accompanying process, which reindeer herders also blamed for losses, was decreasing reindeer tameness, resulting from the social consequences of Soviet and postsoviet socioeconomic transformations, such as long separation from families, workforce scarcity, alcoholism, and, in the last decade, rapid climate change. The use of modern technology, like snowmobiles and caterpillar-tread vehicles in the tundra, and the influx of visitors are further factors. Another dimension is racial and colonial stereotypes grounded in the ideology of social evolution and the dogma of Indigenous cultural authenticity, which represent Sami as premodern and "simple" and their culture as "traditional." Indigenous people are thus permanently caged into marginality, which is also extended to reindeer: if they do not follow prescriptions to modernize, they are "primitive" and "backward," if they use modern methods and technology, they are not "authentic" bearers of Indigenous tradition (Vladimirova, 2014).

Anthropologist Yulian Konstantinov describes the ultimate deterioration and depopulation of reindeer herding in the Kola in 2020, in contrast to the exaggerated official statistics. In his view, the quasi-Soviet bureaucratic organizations of the two surviving enterprises exploit statistics to access regional funds intended for the support of Indigenous people, while in reality they employ only a few herders who interact with reindeer on the tundra infrequently, and have little control over the small number of remaining reindeer (Konstantinov, 2023). Meanwhile, the land of the enterprise Tundra is being allotted for mining without the informed consent of Indigenous and local communities. Nornickel and the state corporation Rosatom, two of Russia's largest industrial companies, in 2022 acquired a license for a large-scale lithium mine amidst Tundra's calving territories near Kol'mozero Lake. The company and the regional authorities stress the new project's promise for regional economic and social development, local community resilience, global sustainable energy transition, and social justice, because Indigenous people have been invited into the planning of the new mine, and its negative impact to traditional economy and culture is transparently discussed and fairly compensated. Such reports conceal the reality that the affected territory will be lost forever for reindeer herding and the Indigenous economy. Already in the early 2000s, Indigenous people claimed that mining companies and authorities were interested, invisible agents in the dismantling of Indigenous reindeer herding. What at the time sounded like a conspiracy theory turned out to be wise and insightful, as I will show below through the analysis of the transformation of Zakaznik Seid"iavvr' through the lens of biopolitics.

Zakaznik Seid"iavvr'

Zakaznik Seid"iavvr' is a Soviet-era protected territory of 17,400 hectares, established by a top-down initiative in the 1980s to preserve fish, bird, and animal species. It was revived in 2003 under the influence of public lobbying, environmental NGOs, and Sámi organizations. Seidozero Lake presents both scenic beauty and a sacred site. The re-established zakaznik aimed to strictly protect and develop the cultural monuments and, most importantly, the traditional environment, lifestyle, and original culture of the Sami people. In order to fulfill its goals, it maintained a more restrictive regime: logging, geological exploration, mining, turf extraction, house building, roads, pipelines, electricity systems, and communications infrastructure were forbidden. Measures to achieve its preservation goals were introduced, as well as those for revival of Indigenous land use, such as reindeer herding, fishing, berry picking, and the infrastructure for these; preserving historical and cultural heritage and revival of handicraft and souvenir production; and ritual activities at religious and historical places.

Part of the territory of the zakaznik, pastures previously used by Tundra enterprise, was leased by a Sami *obshchina* ("kin group" in Russian). The obshchina is envisioned as a way to revive traditional family and communal organization for Indigenous subsistence and self-governance. Nonetheless, due to the heavily bureaucratized procedures of legal registration and land access, few Indigenous people take advantage of this possibility. As with the marginality I described above, this is another instance of contradictory treatment of Indigenous rights: on the one hand, land tenure to secure traditional household economy is promised to Indigenous people, but on the other, is being hindered through complex rules and paperwork that require educated urban knowledge, or an expensive legal counselor. Land can only be given to obshchinas within the borders of the so-called "territories for traditional land use," which have not been marked in Murmansk Region, so obshchinas are forced to participate in public lease auctions on the same terms as commercial organizations. The obshchina was created with a Danish grant, which, together with its use of the historical territory of the Tundra enterprise, created turmoil and hostility to it among Lovozero residents, where Sami are a minority. Tundra, as a successor of the Soviet state enterprise, is imagined as a collective asset for the whole village, like a public good, and the obshchina was thus seen as appropriating public resources for its private enrichment. People questioned the justice of only a few Sami getting rights over zakaznik land and benefiting from its fame and symbolic status by building small-scale tourist businesses. Further contestation occurred when the small number of reindeer obtained through the foreign grant did not build self-sufficient reindeer herding in the obshchina but were let graze in enterprise Tundra grazing lands, while the obshchina members were reported to be organizing trips for tourists in the zakaznik.

Sami did not participate in the government of Seid"iavvr'. But the zakaznik manager, a member of the municipal administration, stated his openness to cooperation to ease tourist access to the obshchina territories, as long as they did not violate the preservation of biodiversity and historical heritage. The obshchina, however, could not afford to pay the land lease after the grant ended.

The failure to revive Sami reindeer herding, the goal of the zakaznik, lies first in bureaucratic hindrances to accessing land on a long-term basis and forming domus relations. Obshchina members, though coming from herding families, lacked direct skills and practical knowledge of herding in this territory, which was also new to the reindeer they obtained. The competitive market principle of securing access to land prevents continuous Indigenous land use and enables the circulation of land and its appropriation by powerful economic stakeholders. Indigenous marginalization results from the colonial knowledge practices around land management, and also through laying blame for the failure of zakaznik objectives on the obshchina rather than zakaznik management.

The municipal authorities took advantage of the lease ending to spread rumors that due to the loss of tradition, Sami are not able to revive herding and silently evaded the formally stated purpose of the zakaznik, increasing tourism instead. After management was moved to the regional administration in Murmansk, taxes to visit and fish in the zakaznik were introduced that could only be paid online, allegedly with the purpose of improving protection and tourist infrastructure. As Sami complain, due to frequent failures in the electronic payment system, they are sometimes refused access to the zakaznik.

The next managerial vision of the zakaznik was the project *"Re-creation of an ancient Sami settlement and the building of a cultural and educational environmental center on its basis,"* which has been discussed since 2012 in the municipal and regional administrations. The project aims to build tourism infrastructure. An "ecological visitor's center" will show Sami's everyday life and performances of traditional rituals. An "ethnographic village" where Sami live will also accept tourists. The village will showcase ancient Sami-style architecture but offer contemporary living standards through technologies promoting environmentally friendly use of resources. Lastly, an Indigenous Educational Centre will hold classes for Sami youth in traditional life skills, and in managing the cultural and educational work of the ecological center.

The applicants, the international ecological fund *Clean Seas*, the Ministry of Natural Resources and Ecology of Murmansk Region, the Directorate of nature conservation, and the Limited Liability Company *Made in Arctic*, state the project goals: (1) including Indigenous people in the struggle against the pollution of the Arctic, (2) creating a showcase of symbiosis between traditional northern spirituality and latest scientific innovations, and (3) harmonizing nature and modern technological economy. In addition, the project aims to contribute to sustainable development and revival of Sami culture by educating Sami, popularizing ecological tourism and infrastructural development through green technology, and finally decreasing human pressure on ecosystems by limiting unorganized tourism and poaching.

The applicants promise that Sami people will be employed in the service sector and education. They intend to teach Indigenous people about ecological sustainability while emphasizing Indigenous culture as an example of harmonious life with nature, where Sami's "genetic memory" should help them "adapt" to contemporary green ecology and "green energy" in the Arctic. The applicants mention reindeer herding, but no infrastructure for it is planned. *Clean Seas* claims in meetings about the project within the Arctic Council that Sami organizations and communities support the project, hiding several open letters and media publications that raise concerns and protests (Shchukin, 2017). No impact evaluation of the project on the protected ecosystems, cultural heritage, and Sami culture has been presented. Indeed, it is being kept secret from local residents while PR companies headed by elite scholars are favored and paid for by the regime, and politicians have been negotiating with international organizations to procure funding for "Sami culture." The project creates the conditions for de-facto appropriation of the zakaznik territory by limiting further its formally stated role in reindeer herding, which will exist solely in storytelling and performances in front of tourists. Its history as Tundra grazing land and the revival of Sami traditional economy are being disconnected from reindeer herding and the place that has constituted and maintained domus. Sami are included in the project and allowed in the area only if they take the role of service staff or tourist spectacle. For this purpose, they need to learn how to reinvigorate their "genetic memory" (authenticity) and "adapt" to contemporary green ecology and "green energy."

With Russia's participation in international environmental cooperation in the Arctic now suspended, the motivation for continued observance of Indigenous rights and interests according to international requirements has weakened. The Government of Murmansk Region, in 2022, announced its new Concept of development of protected natural territories until 2025, under which Seid"iavvr' will be converted to a Nature Park. The new organizational form permits splitting the territory of the zakaznik into several zones: a protected zone, a heritage zone, and a recreational zone. The Nature Park territory is directly managed by state authorities. Within the park also, different forms of ownership are allowed, giving legal grounds for privatization of lands within the previous zakaznik. The authorities stress that the reorganization will increase the park's territory, but as Sami commentators point out, the actual protected territory is considerably diminished. Half of the area is to be converted into a recreational zone, including the Sami sacred sites (Berezhkov, 2022). They have become popular attractions for groups practicing occultism and contemporary paganism, which disrupt Sámi sacred beliefs and leave unwanted traces, but have economic potential.

A third zone in the park is planned for "traditional land use," where mining is also allowed, but no Indigenous use is specified. One of the listed objectives of the planned reorganization of Seid"iavvr' is the "realization of sustainable functioning" of the Lovozero Mining and Processing Plant. The area of Lovozero is well known for its rich supply of rare minerals that are in use today in the energy transition economy, optics, electronics, military industry, and other technologies. The reorganization of the zakaznik thus creates economic opportunities for actors close to the authorities more than ensuring the future preservation of biodiversity and ecosystems.

The justification of the planned reorganization still lists the preservation of traditional lifeways, economy, and original natural environment of Indigenous Sami people, as well as their cultural heritage in the territory. After multiple complaints and addresses to the authorities, public discussion of the planned transformation was organized on short notice, and about 15 Sami residents of Lovozero attended. The critical perspectives of this group were not acknowledged by the authorities, who reported the meeting as successful, and in response to a letter of complaint from local Sami representatives, the Murmansk Region minister of natural resources and ecology referred to legal documents showing that there is no legal requirement for public discussion of such a change (Danilov et al., 2022). Meanwhile, industrial companies and state representatives in international forums like the United Nations Permanent Forum on Indigenous Rights describe this as a success story of corporate responsibility and strong processes of building partnerships with Indigenous communities.

Conclusion

Scholars who search for a decolonial approach to resilience emphasize the central role of localized human perspectives, which include Indigenous epistemologies and socioenvironmental justice (Graybill & Petrov, 2020). In the Russian context, however, local resilience is presented as a domain for top-down regional governance. Meanwhile, governmental top-down interventions that claim to strengthen Indigenous traditional economy and culture provide mechanisms for the practical marginalization of Indigenous peoples' role, practices, and historical traces in the territories of their traditional inhabitance, like Zakaznik Seid"iavvr'. In the process, economic actors like mining companies and tourism obtain legitimate and unchallenged access to such lands and resources. Global Arctic resilience evaluations name Indigenous groups with strong and weak resilience, where resilience and cultural survival are implicitly represented as inherent, "genetic" qualities of ethnic cultures. Russian resilience theory appoints centralized administrations as the main agents of resilience. As the example of Zakaznik Seid"iavvr' shows, state authorities claim intentions and make regulations that should allegedly enhance the revival of Indigenous culture and economy and increase Sami welfare. In reality, such interventions do not provide a place for a thriving human-reindeer relationship but envision Sami culture as a tourist spectacle and reindeer as symbolic characters. As reindeer herders from Tundra have commented, the bureaucratic conditions for the imagined "collective" reindeer herding have changed in the last 30 years in ways that put under question the officially stated centralized policy of support for two surviving inheritors of Soviet state enterprises (Konstantinov, 2023; Vladimirova, 2014). The threats to Kola reindeer herding in general, and to the revival of Sami reindeer herding, meanwhile, are continuously being blamed on Indigenous practitioners.

References

Anderson, D. (2000). *Identity and ecology in Arctic Siberia*. Oxford University Press.

Anderson, D. (2006). Is Siberian Reindeer Herding in crisis? Living with reindeer fifteen years after the end of state socialism. *Nomadic Peoples, 10*(2), 87–104. https://doi.org/10.3167/np.2006.100206

Anderson, D. (2017). Humans and animals in northern regions. *Annual Review of Anthropology, 46*(1), 133–149.

Beach, H. (1981). *Reindeer-herd management in transition: The case of Tuorpon Saameby in Northern Sweden,* Doctoral, Uppsala University.

Beach, H., & Stammler, F. (2006). Human-animal relations in pastoralism. *Nomadic Peoples, 10*(2), 6–30. https://doi.org/10.3167/np.2006.100202

Berezhkov, D. (2022). *Kak Rossiia realizuet princip SPOS na praktike (v Saamskom sele Lovozero proshli obshchestvennye obsuzhdeniia)* [How Russia Implies the Principle of Informed Free Agreement in Practice (Public Discussion Took Place in the Sami Village of Lovozero)]. Indigenous Russia. Retrieved from https://indigenous-russia.com/archives/26244

Carson, M., & Peterson, G. (2016). *Arctic resilience report*. Retrieved from Stockholm.

Cavanagh, C. J. (2014). *Biopolitics, environmental change, and development studies.* Paper presented at the Forum for Development Studies.

Danilov, A., Sovkina, V., & Slupachik, A. (2022). *Otkrytoe pis̕mo po voprosu reorganizaciia gosudarstvennogo kompleksnogo zakaznika regional'nogo znacheniia Seid"iavvr' v odnimennyi prirodnyj park* [Open Letter about the reorganization of the state nature complex Zakaznik of Regional Importance Seid"iavvr' into Nature Park].Indigenous Russia. Retrieved from https://indigenous-russia.com/archives/28683

Foucault, M. (2003). *"Society Must Be Defended": Lectures at the Collège de France, 1975–1976*. Macmillan.

Graybill, J. K., & Petrov, A. N. (2020). *Arctic sustainability, key methodologies and knowledge domains*. Routledge.

Grove, K. (2014). Biopolitics and adaptation: Governing socio-ecological contingency through climate change and disaster studies. *Geography Compass, 8*(3), 198–210.

Konstantinov, Y. (2023). *Power and the people: The state and peripheral communities in the Russian Far North*. Springer Nature.

Krupnik, I. (2000). Reindeer pastoralism in modern Siberia: Research and survival during the time of crash. *Polar Research, 19*(1), 49–56.

Larsen, J., Schweitzer, P., & Petrov, A. (2014). *Arctic social indicators: ASI II: Implementation. In.* Nordic Council of Ministers.

Nenasheva, M. (2023). Resilience of arctic communities: Concept, methodology and research directions. *Social and Economic Development* (51), 223–231.

Nilsson, A., & Larsen, J. (2020). Making regional sense of global sustainable development indicators for the Arctic. *Sustainability, 12*(3), 10—27.

Postanovlenie ob utverzhdenii koncepcii funkcionirovaniia i razvitiia seti osobo okhraniaemykh prirodnykh territorii regional'nogo znacheniia Murmanskoi oblasti do 2025 goda i na perspektivu do 2025 goda N 135-ПП [Decree ratifying the plan for functioning and development of the protected natural territories of regional importance in Murmansk Region until 2025 and in perspective until 2035 N 135-PP]. (2022).

Odintsova, A. (2020). From strategic planning to territorial resilience. *Federalism*(4), 26–41. https://doi.org/10.21686/2073-1051-2020-4-26-41

Reid, J. (2013). Interrogating the neoliberal biopolitics of the sustainable development-resilience nexus. *International Political Sociology, 7*(4), 353–367.

Shchukin, G. K. (2017). *Obrashchenie Aborigen-foruma po voprosu stroitel'stva etnograficheskoi derevni Seid"iavvr'* [Address by Aborigen-forum about the creation of an ethnographic village Seid"iavvr']. Indigenous Russia. Retrieved from https://indigenous-russia.com/archives/26380

Stammler, F. (2010). Animal diversity and its social significance among Arctic pastoralists. In F. Stammler & H. Takakura (Eds.), *Good to eat, good to live with: Nomads and animals in Northern Eurasia and Africa* (pp. 215–243). Center for Northeast Asia Studies, Tohoku University.

Vinogradova, V. (2021). Rayonirovanie Rossii po prirodnym usloviyam zhizni naseleniya s uchetom ekstremal'nykh klimaticheskikh sobytii .[Zoning of Russia according to the natural living conditions of the population considering extreme climatic events]. *Bulletin of the Russian Academy of Sciences. Geographical Series*(1), 5–13. https://doi.org/10.31857/S2587556621010167

Vladimirova, V. (2014). "It is not our reindeer but our politicians that are wild:" Contests over reindeer and categories in the Kola Peninsula Northwestern Russia. *Arctic Anthropology, 51*(1), 24–40.

11

Justice, Resilience, and Land Dispossession in the "Green" Transition: Perspectives from Forest Sámi Communities in Sweden

Johanna Ohlsson and Sara Lind

Introduction

This chapter explores how forest Sámi communities in Sweden are affected by the dynamics of the so-called just and green transition, particularly concerning the forestry sector. By utilizing an inductive approach, we study the issue of Indigenous land dispossession, and the interview data we analyze indicate that this dynamic could be explained by combining the concepts of settler-colonialism (Össbo, 2020; Veracini, 2015; Wolfe, 2006) but also recognition justice (Fraser, 2005) as well as resilience (Berkes et al., 2009).

The United Nations has repeatedly criticized Sweden for failing its human rights commitments, specifically for not respecting the right to land for the Indigenous Sámi communities (OHCHR, 2022; UNHRC

J. Ohlsson (✉) · S. Lind
Department of Human Rights and Democracy, University College Stockholm, Stockholm, Sweden
e-mail: Johanna.ohlsson@ehs.se; Johanna.ohlsson@mrs.lu.se

T. Skillington and A. Setti (eds.), *Climate Change Resilience Across Societal Contexts*, Energy, Climate and the Environment,
https://doi.org/10.1007/978-3-031-95020-9_11

2011, 2016). In addition to international critique, the Sámi Parliament has repeatedly warned that Sámi communities will not be able to continue with traditional livelihood when access to land is not protected, nor respected. One of the most pressing issues is competing interests over land use (Allard & Brännström, 2021). Importantly, the Sámi Parliament has stressed that Forest Sámi communities are especially vulnerable, as they have weaker protection of their lands than mountain Sámi communities due to the intensive Forestry Industry (Marklund, 2020; Sami Parliament, 2017, 2020).

The case of Sweden is particularly interesting as Sweden, as part of the narrative of the Nordic Human Rights Paradox, tends to portray itself as a forerunner in respecting human rights (cf. Vik et al., 2020). Importantly in this context, the largest forest owner in Sweden is the state-owned company Sveaskog, which owns 14% of the Swedish forest areas (Sveaskog, 2023). Yet, Indigenous land rights are impugned in many states, including Sweden (Lind, 2023; Norgaard, 2019; Sami Parliament, 2017), posing a direct challenge to Indigenous communities' abilities to continue their traditional livelihood and culture.

In Sweden, the Sámi communities can only exercise usufruct land rights. As a result, they have limited possibilities of participation in discussions about management and tendering of the land. Earlier studies have indicated that other actors, such as companies (both private and state-owned) wanting to establish mines, wind parks, and infrastructure, often use and modify the land that reindeer herders use, posing significant challenges for the Sámi communities (Österlin & Raitio, 2020).

Forestry management is one of these infringements. Central here is the fact that the Forest Industry has played a vital role in the Swedish economy, for instance, mechanized and industrialized forestry was adopted in the northern regions of Sweden during the 1960s. In the contemporary Swedish context, intensive forestry is being practiced on 90% of the forest cover (Harnesk, 2022).

The Sámi communities in Sweden have organized themselves using different strategies to influence the issue of forestry and land rights. For instance, twenty-nine communities wrote an open letter to the forest industries in 2020, alarming the risks of the current development. They state:

> Not even the reindeer that survived on these lands since the ice age can live in the desert landscape that the forest companies, led by state-owned Sveaskog, create. As indigenous people and practitioners of natural pasture-based reindeer husbandry, we must raise the alarm! (Authors translation). (Lundgren et al., 2020)

In addition, the Sámi communities have been collaborating with various non-governmental organizations (NGOs) such as Amnesty Sápmi (Amnesty Sápmi, 2021) and the Swedish Society for Nature Conservation (Lundberg et al., 2020), trying to get the issue of forestry in the political and media spotlight. There is still a lack of research on the specific topic of forestry and its infringements on Indigenous communities in general and Sámi communities in particular. Previous studies have highlighted the necessity to explore the effects forestry has on reindeer herding for those communities whose livelihood is year-round herding in forested areas (Joona and Joona, 2023; Olofsson, 2022; Vihervaara et al., 2009), which justifies our starting point.

The Sámi Communities and Indigenous Land

As stated by Joona and Joona (2023), people have always been keen to find and discover new areas and resources. Many associate this development with overseas excursions and occupation of colonies. In the Scandinavian—or Fennoscandian—context, Denmark, Sweden, and Russia were instead centering on the areas of northern Fennoscandia, known as Sápmi. However, Sápmi was already inhabited by the Sámi people, the only Indigenous people within the EU. Sámi communities have resided across Scandinavia (Sweden, Finland, Norway) and Russia long before the Swedes, Danes, and Russians colonized their lands (Joona & Joona, 2023, p 124f).

Over time, Sámi communities became part of the Swedish legal system and began paying taxes. For some time, the Swedish state did recognize and respect the Sámi communities' territories, but this shifted during the fifteenth and sixteenth centuries when the Swedish state started to claim ownership over traditional Sámi land (Marklund, 2020). In 1749, a regulation called "*Lappmarksreglementet*" was adopted, leading to more widespread colonization as this initiative granted privileges to settlers who resided or relocated to northern Sweden. Consequently, many Forest Sámi communities started to lose more of their land to Settlers and were pushed further north. Some Forest Sámi communities even ceased to exist. In this, the "cultivation border," i.e., the border separating arable from non-arable land, played a significant role. The cultivation border was both a physical border, distinguishing various kinds of land, but it was also a legal regulation. The border was drawn in 1753, and it was determined that Sámi communities could live and practice traditional livelihoods on the west and north sides of the border. Sámi communities who lived on the east and southern sides of the border, which in practice were mainly Forest Sámi communities, were forced to move or abandon their traditional way of living (Össbo, 2020).

The Swedish state has expanded its control over Sámi territory, partly through developments within the national legal system. The Swedish state justified this by diminishing Forest Samis and portraying them as incapable of managing forests and land (Marklund, 2020). The contemporary regulation of Sámi land rights is governed by the Reindeer Herding Act (RHA, *Rennäringslag* 1071: 437) from 1979 and amended in 1993. It is stated that Sámi communities have usufruct land rights based on immemorial prescription. Importantly though, note that they lack ownership rights. Strömgren (2015, pp. 95–97) has argued that the RHA is the only law regulating Sámi people's land rights. Strömgren states that when RHA first was enacted in 1886, it was formed to regulate the relationship between settled farmers and reindeer herders on reindeer winter grazing land. The law did not aim to manage the Sámi communities' primary living areas. It is, therefore, problematic that RHA has come to function as a codification of exhaustive Sámi rights to land.

The Sámi communities serve as an economic association where membership is required to engage in reindeer herding. These communities and villages are categorized as either Mountain villages, which conduct reindeer herding in highlands during summer and move to the forest in the winter, or Forest villages, which practice year-round reindeer herding in the woods. The historical and contemporary conditions and experiences of Forest Sámi communities have received less scholarly attention (Marklund, 2020; Olofsson, 2022). This division has been explained as rooted in the Swedish historical classification of Sámi communities. According to Össbo (2020), the state categorized Sámi communities in the past based on their nomadic lifestyle. The Mountain Sámi communities, who lived as nomads to a greater extent, were considered the "authentic Samis," while the Forest Samis were seen as a deviation (Össbo, 2020).

In addition to this historical development, the narrative of Sámi communities' approach and understanding of nature cuts across these changes, relating to pre-existing environmental knowledge. Embracing a holistic understanding of nature as deeply intertwined with us as beings, nature has a significant role and influence on the social life of Indigenous peoples in various parts of the world. State initiatives have been shown to have impacted physical and mental health, causing loss of traditional knowledge, and weakening relationships in the Indigenous community (see Norgaard, 2019 for the US; Össbo, 2020 for Sweden). Sámi as well as other Indigenous communities' traditional ecological knowledge and sustainable land management practices have been established as a key resource in the just and green transition (Allard, 2024), but compliance and implementation are still lacking.

Relevant Legal Instruments in the Swedish Context

Taking as a starting point the understudied dynamics and experiences of the year-round forest reindeer herders, our field research explored the relevance of their experiences with Sámi interview participants. A Human Rights-Based Approach (HRBA) was adopted to understand how human rights implementation is affected by factors like history, culture, and economy (McConnell & Smith, 2018, pp. 9–12). Human rights are seen as a normative standard for human dignity codified by international human rights law, being realized through domestic laws, policies, and programs. As is common within an HRBA, we positioned our study in the common disparity between international human rights law (IHRL) and people's lived experiences. To identify underlying issues in the implementation of human rights, we followed the approach of Landman and Carvalho (2010) examining two out of their three distinct levels. We first examine rights in principle (level 1), i.e., a doctrinal approach, and a legal analysis of international and national law, and later move on to rights in practice (level 3).

In tandem with the domestic legal and political development on Indigenous issues in Sweden described in the previous section, there have been significant international advancements (cf. Shelton & Guzmán Duque, 2024, p. 13ff). This is central from a human rights perspective. Sweden has a dualist legal system and has ratified the International Covenant on Civil and Political Rights (ICCPR), the International Covenant on Economic, Social, and Cultural Rights (ICESCR), and most of the other conventions. Of relevance here is also the Convention on the Elimination of All Forms of Racial Discrimination (CERD), as the CERD committee regularly has posed criticism regarding the states' lack of compliance with Indigenous rights (cf. Shelton & Guzmán Duque, 2024, p. 14ff). In addition, the European Convention on Human Rights (ECHR) has been incorporated into Swedish law since 1995, and the Convention on the Rights of the Child (CRC) since 2020. However, the issue of Indigenous rights has for long politically been ignored in Sweden, and Sweden has not ratified ILO 169. In the legal analysis, the importance of ILO 169 became clear and it is, therefore, included as material of this study as soft law.

The other material used for the legal analysis consists of UN conventions and declarations such as UNDRIP and relevant articles in ICCPR, ECHR, and CERD. These were selected over regional ones as they offer the most comprehensive legal framework of Indigenous rights (OHCHR, n.d). Academic articles and relevant verdicts have been consulted to see how the laws are used and interpreted. The domestic laws analyzed are RHA and the Forestry Act, which show the difference in rights between the Sámi and the majority population landowners. The main issue within the Swedish legal framework is that the Sámi people only have usufruct rights. The landowners, in this case forest companies, have the right to decide how they want to use their land. The Forestry Act allows forestry methods that harm reindeer herding such as clear-cutting, and the use of the tree species contorta, which has repeatedly lowered the lowest felling age.

In addition, the *Act (2022:66) On Consultation on Issues Concerning the Sámi People* (Consultation Order) is also included as it directs Sámi influence over matters that may affect them. Centrally, however, before a cutting reindeer communities must be consulted, but no guidelines are given as to how the companies should consider the reindeer communities' opinions. Similarly, the Consultation Order has no provisions regarding how Sámi communities' positions must be considered in the decisions that follow a consultation.

The "Green" Transition and Forestry in Sweden

Sweden's landscape is dominated by forests, which cover over half of the country's total land area, primarily consisting of coniferous trees. These forests provide diverse habitats for a wide range of species. Beyond their ecological significance, they are also a key source of renewable raw materials. The Swedish state has since 1647 regulated the use of the forest through national laws, limiting how Forest Sámi communities can use and live by their traditional land. However, previous studies have stressed the need for more research to better understand how forestry affects Forest Sámi communities (Lind, 2023; Olofsson, 2022; Össbo,2022).

Current research on Swedish forestry and the green transition is somewhat divided. While some see the practice of reforestation as ensuring sustainable use of the forest, others argue that the high rate of forest felling increases carbon dioxide emissions and reduces biological diversity (Lundmark et al., 2017; Svensson et al., 2020). Another issue is that industrial actors have influenced some forest research programs and that these tend to steer research to focus on industrial productivity rather than sustainability (Andersson & Westholm, 2018). This latter is highly relevant for Forest Sámi herders as they depend on a healthy and biodiverse forest and constitutes topics that necessitate more research.

Previous research on forestry and Sámi rights has primarily focused on lichen pastures used by the reindeer during winter. Reindeers rely on terrestrial lichens for food, but when ice covers the ground, they turn to arboreal lichens. Clear-cutting and intense site preparation by the forest industry have reduced these lichen resources, with a severe impact on reindeer herding (Kivinen et al., 2010). Hernesk (2022) highlights how climate change worsens this issue, as ice crusts become more common, further limiting access to terrestrial lichen. Despite a decline in pasture quality, reindeer populations remain stable due to costly herding techniques like supplementary feeding, which are unsustainable and threaten forest wellbeing (Uboni et al., 2020).

Research on lichen access sheds light on how the forest industry impacts reindeer herders, but few studies have focused on the experiences of forest reindeer herders who spend the entire year in the forest. This is where our study contributes to the state of the art. Others, such as Vihervaara et al. (2009), conclude that practices that alter the land cover significantly, such as Forest Industry or road networks, had the most substantial negative impact on reindeer forest communities in Finland as well as ecosystem services.

To best understand the effects of forestry on Sámi practices and livelihood, this must be done in a context that also considers other infringements. Österlin and Raitio (2020) have investigated the cumulative effects of encroachments on Sámi districts affected by multiple competing land uses, such as hydropower, mining, forestry, and infrastructure development. What is clear from their study is that the lack of integrated spatial planning in reindeer herding areas is generating new challenges for Indigenous rights. This is a problem as the task of conducting integrated landscape planning and defending their districts through various consultations with different industries becomes close to unbearable for the affected Sámi communities (Lind, 2023). Taken together, these developments cumulatively undermine Indigenous resilience by restricting access to traditional lands, depleting vital ecological resources, and imposing additional economic and bureaucratic burdens on Forest Sámi communities. Further, the expansion of industrial forestry, coupled with other competing land uses, has fragmented reindeer pastures and eroded the biodiversity necessary for sustainable herding. Indigenous resilience is in this context referred to the reindeer heeders' ability to maintain traditional herding practices, adapt to environmental changes, and sustain their livelihoods despite external pressures. The lack of integrated spatial planning further exacerbates these pressures, forcing Sámi herders into costly adaptations like supplementary feeding, which are neither economically nor environmentally sustainable. Ultimately, the cumulative effect of these encroachments threatens not only Sámi livelihoods but also their cultural traditions and ability to exercise self-determination over their ancestral lands.

Forestry in Sápmi—Voices from Sámi Communities in Sweden

We also explore rights in practice (level 3 in the Landman and Carvahlo 2010 approach) via the co-generation of data in semi-structured interviews. These two levels have influenced each other, firstly as the legal analysis provided a framework for creating the interview guide and analyzing and structuring the interview results. Secondly, the interview results helped reveal some parts of the legal framework as particularly important. The conversations that have generated the interview material have been carried out with representatives from five different forest Sámi communities in Sweden. Eight conversations have been generated, and the participants have been recruited through selective criteria and snowball sampling. The eight participants were all males (explained by the selection methods), aged between 26 and 68 years. For future studies, it would be interesting to also generate data through conversations with women from forest Sámi communities. Interestingly, there is no significant variation in the views or perspectives regarding generational aspects.

Six of the participants are working full-time as reindeer herders, while two are combining reindeer herding with representation in the Sámi Parliament. The conversations have centered on land-use regulations' focus on forestry management, an infringement shared by all forest Sámi communities, as well as the implications of the forestry industry for the communities. As stated in the chapter's introduction, forestry is one of many infringements the forest Sámi communities are facing, and it is faced on a large scale. As one participant puts it:

> ... Perhaps the biggest exploiter of all is forestry. ... It's an intrusion with effects just about everywhere; that's how it is. A mine can have a huge impact but in a local place. Forestry is on every single square meter.

The consequences that are mentioned are that pastureland shrinks, but also that there is a high rate of tree felling and clear-cutting. In tandem, there is a common trend to pursue contorta, i.e., creating plantations rather than creating space for nature. These aspects have direct negative effects on the reindeer herding practices for forest Sámi communities. One participant said:

> ... It doesn't work if you only have one type of forest all the way, along your migration route. Today, it's like one plantation with a young forest and then one with a 60-year-old forest; it doesn't work ... you need to have nature. In which the reindeer does not feel at home, it is hard to find pasture, and the lichen does not spread.

This statement can—in the light of the United Nations Biodiversity Convention—be seen as centrally connecting current forestry practices with the conservation of nature as well as the respect for Indigenous peoples' rights. Here it becomes clear that corporate interests are conflicting with rights, which also affects economic structures. The participants describe how forestry practices have pushed reindeer herding into a downward spiral—the pastureland shrinks, the herders have adopted methods that increase the cost, and one needs a greater herd to make a living, but the pastureland cannot support more reindeer, many herders leave their vocation. This is in great contrast to the growth of the forest industry. One participant summarizes this neatly: "*... Forestry favors consumption, it favors society's strive to increase, and grow and become bigger constantly, and above all, it favors financial gain.*" In addition, social structures of state-Indigenous but also majority-minority relations are reinforcing the economic growth of the forest companies. One participant explains: "*... The biggest obstacle is that the majority* *society always wants more. Everyone just wants to be more comfortable and raise the standard of living. Nobody wants to go back. So that is the biggest challenge, how to manage to stop this wheel."*

The forest Sámi communities' main tool to impact forest management is consultations with the forest companies. These consultations are time-consuming as the Sámi communities must attend consultations regarding each possible infringement. While forest industries must consult with the Sámi communities, there are no guidelines on how their interests should be considered neither in the Forest Act nor in the Consultation Order, one participant stated "… *that's a dilemma, having a law, but you can never use it*." The Sámi communities are left with toothless laws to defend themselves against an industry that slowly eliminates their land and makes it impossible to uphold traditional approaches to living and livelihoods.

This, we argue, taps into a logic of elimination. The disposal of land has shifted from the Indigenous people to the state and its companies and majority citizens. This is done through structures that continue to reinforce this logic, contributing to making current forestry practices anything but sustainable. In contrast, several participants highlighted how their way of living is in its essence sustainable and results in resilience toward climate change. One of them puts it like this:

> ... I actually think that Sweden should consider itself lucky to have an Indigenous people who fought back against the forest invasion and its pace. ... I think it would have looked completely different if there hadn't been someone complaining that "You can't continue at this rate!" Then it would have been just free to cut and cut in at an even faster pace.

Another participant connects their fight to the global Indigenous community and states:

> ... We must realize that 5-6 percent of the world's population is Indigenous, and we possess 25 percent of the earth's surface, on which there is 80 percent of the remaining biodiversity -and this is extremely valuable for having an inherent power against global warming.

This quotation helps pinpoint the significant role Indigenous groups have in the transformation toward a more sustainable future. It also speaks to how climate change is expanding social inequalities between different societal cohorts and increasing vulnerability to experiences of harm (Skillington & Setti, 2025). In addition, when elaborating on the effects climate change, one could connect to the discussions the participants had on the heightened stress, anxiety, grief, and marginalization, but also the lack of pastureland as well as shrinking land in relation to the changes they experience in the forest. Centrally, the reindeer herding participants of our study are among those who encounter the impacts of climate change in more than one way. It is an issue of a loss of land and livelihood, physical and mental health, security of place, biodiversity, and cultural heritage, making it both individual and collective experiences of harm.

On the other hand, and in the context of the state-Indigenous but also the majority-minority dynamic, the act of reindeer herding seems to sometimes be seen as an act of resistance toward marginalization. Several participants described an ongoing fight against the forestry processes that are happening under the auspices of the "just and green" transition, highlighting how the struggle is carried through in courts, in consultations, and by continuing to practice reindeer herding. Here, the act of herding is seen as a form of resistance, and nature is a source of resilience. We see a clear parallel in the resistance toward settler-colonialism and resistance toward climate change, also laying out the necessity to add a power-sensitive analysis to processes against climate change. If not, the so-called just and green transition risks ending up yet another tool for colonialist practices, hierarchies, and dominance (Harnesk, 2022; Lundgren et al., 2020; Österlin & Raitio, 2020). This tendency, we argue, goes in line with what Skillington and Setti describe as a motivational power of heritage and communal resources (Skillington & Setti, 2025).

Conclusion: Resistance and Advocacy

Sámi resistance to so-called green projects has taken the form of concrete resistance and litigation processes, as well as collaborations between Indigenous groups, environmental NGOs, and international organizations. What is clear from our study is that the dynamic between Sámi communities and the forestry sector and therefore also the Swedish state is best described as the experience of non-recognition, injustices, human rights violations, and land dispossession.

We have worked inductively in this study, and the findings from the interview data speak toward themes in recognitional justice, following Nancy Fraser's three Rs—representation, redistribution, and recognition. As elaborated on earlier, the Sámi communities in Sweden have limited possibilities to representation due to the usufruct standing of their land rights.

So, in ending this chapter, what lessons can be drawn that are relevant to the broader global context of Indigenous land rights and the green transition? We argue that the eco-social relationship in the Swedish forest, historically as well as in temporary social structures, is tainted by non-recognition, injustices, and human rights violations. This means that governments, companies as well as international bodies must work harder to reconcile Indigenous rights with the urgent need for climate action. As the chapter has shown, this process is filled with interests that collide with rights. But importantly, Indigenous rights cannot be reduced to another set of interests, charity, or privileges. The fact that they are both internationally and nationally recognized rights implies that they are legitimate and non-negotiable claims Indigenous communities have toward the state.

In this chapter, we show that the disposal of land affects the social, economic, and political structures within a settler state, which also negatively influences minorities such as Indigenous communities. The main aim of the just transition is to ensure a fair and equitable process of moving away from fossil fuels and toward the adoption of renewable and low-carbon technologies, disrupting, reconfiguring, and usurping the prevailing carbon-intensive global regime to enhance justice and sustainability (McCauley, 2023). To foster this transition toward these goals, integrating Indigenous rights into the processes is pivotal. Without respect for Indigenous rights, and not taking consultation orders and consent processes seriously, there cannot be a truly just and green transition.

Competing Interests and Ethics Approval:

Sara Lind Conducted the data collection during her course 5MR554 Degree Project Human Rights and Democracy, Spring 2023 at University College Stockholm. All research ethics procedures have been adhered to, and informed consent was obtained from the individual participants.

Johanna Ohlsson has received support from a research grant from the Finnish Research Council for this work (REBOUND).

References

Allard, C. (2024). Sámi rights in the sustainable transition—concluding remarks. In D. Cambou & O. Ravna (eds.), *The significance of Sámi rights: law, justice, and sustainability for the indigenous Sámi in the Nordic countries* (pp.183–202). Routledge.

Allard, C., & Brännström, M. (2021). Girjas Reindeer Herding Community v. Sweden: Analysing the Merits of the Girjas Case. *Arctic Review on Law and Politics, 12*(0), 56–79.

Amnesty Sápmi. (2021). *Infekterad debatt mellan Muonio sameby och Sveaskog*. https://amnestysapmi.se/infekterad-debatt-mellan-muonio-sameby-och-sveaskog-2/. Accessed May 01, 2023.

Andersson, J., & Westholm, E. (2018). Closing the future: Environmental research and the management of conflicting future value orders. *Science, Technology, and Human Values, 44*(2). https://doi.org/10.1177/016224391879

Berkes, F., Colding, J., & Folke, C. (Eds.). (2009). *Navigating social-ecological systems: Building resilience for complexity and change*. Cambridge University Press.

Fraser, N. (2005). Mapping the feminist imagination: From redistribution to recognition to representation. *Constellations, 12*(3), 295–307.

Harnesk, D. (2022). The decreasing availability of reindeer forage in boreal forests during snow cover periods: A Sámi pastoral landscape perspective in Sweden. *Ambio, 51*, 2508–2523.

Joona, T., & Joona, J. (2023). The complex relationship between Forest Sámi and the Finnish State. In W. Donnelly & J. Ohlsson (Eds.), *Arctic justice: Environment, society and governance*. Bristol University Press.

Kivinen, S., Moen, J., Berg, A., & Eriksson, Å. (2010). Effects of modern forest management on winter grazing resources for reindeer in Sweden. *Ambio, 39*, 269–278.

Landman, T., & Carvalho, E. (Eds.). (2010). *Measuring human rights*. Routledge.

Lind, S. (2023). *Swedish settler-colonialism in the forest: Forest Samis' rights and land disposal*. University College Stockholm.

Lundgren, P., et al. (2020, November 26). Skogsbolagen skövlar våra renbetesmarker. 29 samebyar: Att verksamheten skulle vara klimatanpassad och rättvis är en bluff. *Aftonbladet*. https://www.aftonbladet.se/debatt/a/X891bo/skogsbolagen-skovlar-vara-renbetesmarker. Accessed September 08, 2024.

Lundmark, H., Josefsson, T., & Östlund, L. (2017). The introduction of modern forest management and clear-cutting in Sweden. Ridö State Forest 1832–2014. *European Journal of Forest Research, 136*, 269–285.

Lundberg, L., Nilsson, J., Lidström, M., & Mildh, B. (2020, June 10). *Baksidan av de vackra formuleringarna om hållbart skogsbruk!*. https://norrbotten.naturskyddsforeningen.se/baksidan-av-de-vackra-formuleringarna-om-hallbart-skogsbruk/ Naturskyddsföreningen, Norrbottens län. Accessed September 08, 2024.

McCauley, D. (2023). Just transitions. In J. Ohlsson & S. Przybylinski (Eds.), *Theorising justice: A primer for social scientists*. Bristol University Press

McConnell, L., & Smith, R. (Eds.). (2018). *Research methods in human rights*. Routledge.

Marklund, B. 2020. Skogssamernas historia, ett oskrivet kapitel? In A. Össbo, B. Marklund, Lena M. Nilsson & K. Stoor (Eds.), *Skogssamisk Vilja: En jubileumsantologi om skriften "Dat läh mijen situd", Karin Stenberg och skogssamisk historia och nutid* (pp. 117–144). Várdduo, Centrum för samisk forskning.

Norgaard, K. (2019). *Salmon and acorns feed our people: Colonialism, nature, and social action*. Rutgers University Press.

OHCHR. (2022). Joint communication from special procedures. AL SWE 2/2022. https://www.amnesty.se/documents/232/AL_SWE_03.02.2022_2.2022.pdf. Accessed September 08, 2024.

Olofsson, S. (2022). Att få utrymme i skogen: Staten och skogssamerna i mellersta Sverige under 1700-talet. In C. Johansson, H.-E. Nilsson, P. Öhman, & B-G. Jonsson (Eds), *Skogens Värden - forskares reflektioner.* Mittuniversitetet. Accessed February 06, 2022.

Össbo, Å. (2020). Från lappmarksplakat till anläggarsamhällen: Svensk bosättarkolonialism gentemot Sápmi. *Historisk Tidskrift, 140*(3), 420–443. Accessed Feburary 02, 2023.

Össbo, Å. (2022). Hydropower company sites; a study of Swedish settler colonialism. *Settler Colonial Studies, 13*, 115–132.

Österlin, C., & Raitio, K. (2020). Fragmented landscapes and planscapes—The double pressure of increasing natural resource exploitation on indigenous Sámi Lands in Northern Sweden. *Resources, 9*(9), 104.

Sámediggi [Sami Parliament]. (2017, November 14). Comments of the Swedish Sami Parliament regarding Sweden's 22nd and 23rd periodic report to the committee on the elimination of racial discrimination. Dnr. 1.2.7–2017–1532. Accessed Feburary 22, 2023.

Sámediggi [Sami Parlament]. (2020). Uttalande om Skogsbruket. https://www.sametinget.se/155898. Accessed September 08, 2024.

Shelton, D., & Duque, F. G. (2024). *Advanced introduction to indigenous human rights*. Edward Elgar.

Skillington, T., & Setti, A. (eds.) (2025). Introduction. In T. Skillington & A. Setti (Eds.), *Climate change resilience across societal contexts: Challenges, potentials and visions for the future*. Palgrave Macmillan

Strömgren, J. (2015). The Swedish state's legacy of Sami rights codified in 1886. In C. Allard & S. Funderud Skogvang (Eds.), *Indigenous rights in scandinavia: Autonomous Sami Law*(pp. 95–110). Routledge.

Sveaskog. (2023). *Det här är Sveaskog*. https://www.sveaskog.se/om-sveaskog/det-har-ar-sveaskog/. Accessed May 03, 2023.

Svensson, J., Bubnicki, J., Jonsson, B., Andersson, J., & Mikusiński, G. (2020). Conservation significance of intact forest landscapes in the Scandinavian Mountains Green. *Landscape Ecology, 35*(7), 2113–2131.

Uboni, A., Åhman, B., & Moen, J. (2020). Can management buffer pasture loss and fragmentation for Sami reindeer herding in Sweden? *Pastoralism, 10*, 23.

UNHRC. (2011, June 06). *Report of the Special Rapporteur on the rights of indigenous peoples, James Anaya. The situation of the Sami people in the Sápmi region of Norway, Sweden, and Finland*, A/HRC/18/35/Add.2. Accessed February 13, 2023.

UNHRC. (2016, August 09–16). *Report of the special rapporteur on the rights of indigenous peoples on the human rights situation of the Sami people in the Sápmi region of Norway, Sweden, and Finland*, A/HRC/33/42/Add.3. Accessed February 22, 2023.

United Nations. (n.d). United Nations declaration on the rights of indigenous peoples. https://www.un.org/development/desa/indige nouspeoples/declaration-on-the-rights-of-indigenous-peoples.html. Accessed February 22, 2023.

Veracini, L. (2015). *The settler colonial present*. Palgrave Macmillan.

Vihervaara, P., Kumpula, T., Tanskanen, A., & Burkhard, B. (2009). Ecosystem services–A tool for sustainable management of human–environment systems. Case study Finnish Forest Lapland. *Ecological Complexity.*, *7*, 410–420.

Vik Hagtvedt, H., Hanne, S. L. B., Jensen, L. L., & Strang, J. (2020). Histories of human rights in the Nordic countries. *Nordic Histories of Human Rights* (pp. 1–13). Routledge.

Wolfe, P. (2006). Settler colonialism and the elimination of the Native. *Journal of Genocide Research, 8*(4), 387–409.

12

Examining Values and Resilience: Climate Change, SAR and Northern Communities

Corine Wood-Donnelly

Introduction

In the northern latitudes, environmental change is now evident in timescales not usually noticed by the inhabitants who live in these cold and remote places. Arctic peoples have over time developed technologies and practices that have enabled life and survival in the far north, traditions passed from generation to generation. Today, the scientific data tells us that global warming is increasing at four times faster than elsewhere on earth (Rantanen et al., 2022). In the Arctic, this warming brings with it environmental state change (e.g. sea ice melt, permafrost thaw) that is at odds with the ways of seeing, being and doing that have underpinned resilience for northern communities. In parallel to the environmental changes, we can point towards governance arrangements that compound the variables for resilience, in particular when responsibility is siloed into

C. Wood-Donnelly (✉)
Nord University, Bodø, Norway
e-mail: c.wood-donnelly@nord.no

T. Skillington and A. Setti (eds.), *Climate Change Resilience Across Societal Contexts*, Energy, Climate and the Environment,
https://doi.org/10.1007/978-3-031-95020-9_12

regulatory frameworks, a rigidity which can have dire consequences for the effectiveness of Arctic search and rescue (SAR).

Within communities, SAR can be understood as both an enabler and an indicator of resilience, demonstrating a community's capacity to respond to and recover from emergencies. While there are many definitions of community resilience found in the literature (see Koliou et al., 2018), a synthesised set of indicators for community resilience can include dynamics such as absorptive capacity, elasticity, self-organisation and adaptation, capacity for resource mobilisation, social capital and institutional capacity, multi-hazard resilience, ecosystem stability and regeneration, risk reduction and preparedness and sustainability for long-term viability. Community disaster resilience is also seen as embedded in a specific "cultural and social context" (Cox, 2015 in Kikkert & Lackenbauer, 2021, p. 50). With long histories of survival, northern communities will have been their own "safety net" in times of disaster and have also acted as responders for the disasters of inexperienced Arctic travellers.

Resilience is a concept that is gaining prominence in various fields and is increasingly used in reference to Arctic sustainability (e.g. Graybill & Petrov, 2020; Caymaz et al., 2023). Resilience is generally understood as a system's ability to maintain relationships within a system and functionality amid changes and challenges (Barasa et al., 2018; Mayar et al., 2022). Underlying the elasticity that is inherent in the concept of resilience, there is a growing body of literature that reflects on the role of cultural values in their capacity to hinder or enable the adaptive capacity required for resilience to exist (e.g. Bowland & Lawrence, 2020; Drzewiecki et al., 2020; Posch et al., 2019; Stark, 2014), with values suggested to be as critical as other social indicators—such as demographics, economy or education.

The International Panel on Climate Change has noted the critical need to take values into account as impacting resilience towards climate change and "whether or not communities adapt to climate change" (Cardona, 2012, p. 84). The report notes that values inform the underlying frames, worldview and attitudes, "notions of agency and responsibility" (Cardona, 2012, p. 95) and "judgements about reliability, relevance and weight of competing knowledge claims" (Jasanoff &

Wynne in Cardona, 2012, p. 95). In other research, it is noted that "hope arises from a sense of moral and social order embodied in the expression of key cultural values" (Eggerman and Panter-Brick, 2010, p. 71) and are critical in "reinforcing livelihood resilience" which contribute to "enabling different forms of action", (Daskon, 2010, p. 1097). In sum, although individuals' and communities' set of core human values, or values systems, provide a mechanism to reflect on adaptive capacity with value systems being "instructive for guiding how.

humans interact with and adapt to our natural environment" (McMillen et al., 2014, p. 57), the role of core values in enabling resilience is often overlooked. This research aims to contribute to this gap in the literature on resilience and adaptive capacity.

With the role of values in resilience and adaptive capacity towards climate change still considered as a frontier of scientific research across many disciplines and fields—from sociological research in social capital to disaster preparedness studies—there is still much to be explored, in particular, regarding the Arctic region. This paper contributes to this limited area of research by examining values operative in Arctic communities' search and rescue efforts (SAR) and governance frameworks regulating the same, against a background of deepening climate change. To examine how values, legal frameworks and community practices influence resilience capacity, the discussion below draws on forty interviews with search and rescue volunteers and practitioners from the Canadian and Norwegian Arctic in 2021–2022, conducted as part of the JUSTNORTH project, funded by the EU Commission, as well as detailed document analysis of reports on legal frameworks governing the provision for SAR and state responsibilities. The specific focus of this document analysis was on variations and disruptions emerging from the values and value systems present within the documents under review.

Why Arctic SAR?

Arctic SAR's legal, technical and social infrastructure is situated in a particularly unique context in the landscape of provision for response, recovery and safety for Arctic communities. Similar to other small communities, be it geographic communities or small organisations, capacity is often contingent on role multiplexity. Another way of phrasing this is that frequently, you'll hear that some people wear many "different hats". This places individuals at many different nodes within a social ecosystem. In Arctic communities, one could find that a person is simultaneously filling two or more important functions in a small community, in addition to being a key actor in SAR, potentially putting strain on the community's response system, but also potentially enriching the community's social ecosystem.

Another factor is that Arctic SAR is contextualised by the potential for harsh climate or difficult weather conditions. The moments when SAR is needed are likely to be the moments when the conditions are at their worst given that "temperature and ice conditions are predictive of the probability of a search and rescue taking place on a given day" (Clark et al., 2017, p. 47). When added together with limited infrastructure—such as ordinary civil infrastructure (e.g. roads or mobile phone masts) and other resources, we can understand that the remoteness of Arctic communities creates the necessary conditions for self-reliance, capacity and the need for "planning for the deployment of limited search and rescue assets and capabilities in remote regions" (Chircop et al., 2020, p. 6). This makes the focus on resilience all the more pertinent for the consideration of SAR in Arctic communities.

In some parts of the circumpolar north, the presence and necessity for sustaining Arctic communities is intrinsically connected with the sovereignty of the state and the responsibility in the social contract to ensure security for citizens. Within these particular territories, Arctic states have relied on the "implementation of the political technologies and systemic structures of the state: government administration, colonialism, resource management, taxation…" (Wood-Donnelly, 2019, p. 5) to integrate these peripheries into their national accounts. Responsibility for SAR appears among the many political technologies used to demonstrate their legitimacy for governance and authority over and within this space (Wood-Donnelly, 2024). This nexus between SAR and the Arctic places a distinct lens on how responsibility for the provision of SAR can be considered in the context of how it is exercised "on the ground" in Arctic communities.

The effects of climate change have caused a shift in the operational landscape for SAR in the Arctic. Not only do more ships and tourists visit and transit the region (Arctic Council, 2024), but the everyday environment for Arctic residents, such as ice thickness or seasonal weather patterns, is increasingly volatile. These combined impacts place additional pressure on states to maintain reliable response systems for a widening range of potential incidents (Malik et al., 2024), but critically, also place stress on local communities that may not always be addressed in policy and planning efforts.

This connection between the SAR, the state and the Arctic also exists in spaces that are extra-territorial in international waters. For example, the Arctic SAR Treaty, agreed in 2011, "reinforce the sectoral sovereignty claims…and also largely follow existing bilateral treaty lines" (Wood-Donnelly, 2013, p. 306). This repetitive overlay creates continuity over time of state understandings of spatial claims and preferences to the Arctic. In Arctic offshore waters, SAR provision and cooperation are largely implemented through military resources (e.g. vessels, and dual-use technologies). However, these assets are also deployed in inshore contexts, and in some cases, on land (e.g. Arctic Rangers or the Canada Coast Guard Auxiliary), particularly in SAR exercises, intended to improve the interoperability of these different dimensions.

Both the Arctic Rangers and the Canada Coast Guard Auxiliary (CCGA) are notable for the role that both local and Indigenous knowledge play in the provision of societal safety. With the Rangers conducting patrols in remote regions (Government of Canada 2022) and the CCGA providing equipment and training for a new generation of first responders, these state-sponsored efforts to integrate response capacity at the local level and link together two different forms of knowledge. This can be considered as contributing to resilience by enabling "different types of knowledge and practices of societies accumulated through a continuous interaction with their natural surrounding" (Cuaton & Su, 2020, p. 2). They provide a strong link between traditional livelihoods and ways of being, knowing and doing together with the contemporary need for SAR capability in Arctic communities. In this, participation in SAR not only provides an essential community service but also creates a critical practice to sustain and transfer knowledge of place across generations and to integrate and unite with newcomers over common goals. These nodes of connection and knowledge transfer are essential for reinforcing and sustaining resilience in Arctic communities.

The Values Landscape

Although research on values is not new from a philosophical perspective, more recently, the role of values has emerged as an area of critical inquiry in resilience, adaptation and other forms of political participation. In 2012, a report from the International Panel on Climate Change suggested that values are "important in influencing action on climate change", are critical in "describing what is desirable and preferable" and that there is a need to "recognise [e] and reconcil[e] conflicting values" (O'Brien et al., 2012, p. 446). In the context of SAR, a study of values is important for understanding alignment between different groups, or other elements (i.e. legal frameworks or organisational structures), but also for understanding why volunteers are motivated to participate, particularly in connection to community belonging and shared cultural or organisational perspectives. Ultimately, alignment, or misalignment, of values can indicate the capacity for resilience.

In the process of examining values and resilience connected to climate change, SAR and Northern communities, there are several cultural value sets and resilience factors to be considered. This includes the different values emerging in three different dimensions—SAR legal documents, SAR volunteers and also values from Indigenous and local communities in the Arctic (see Table 12.1). Together, this creates the values landscape in which resilience can be contextualised. The value set was assembled from a survey of a variety of values documents, including the Rokeach Survey (Rokeach, 1973) and Schwartz's work on universals (Schwartz, 1992), and core legal texts, such as the Paris Agreement and the Rio Declaration 1992. These values identified for this discussion emerged from a document and interview coding process, using the set of values identified in the values document survey, many of which represent the "minimum requirement of good life" (Kekes, 1993, p. 204).

Table 12.1 Comparative value sets

	Legal frameworks	SAR volunteers	INUIT community
Values	Human security	Human security	Respect and care for people
	Provisioning	Provisioning	Serving the community
	–	Connectivity	Collaborative teamwork
	Environmental protection	–	Respect and care for the environment
		Capability	Skill development through practice
		Wisdom	Innovation and resourcefulness
		Connectivity	Openness and inclusivity
	Efficiency	Responsiveness	–
	Responsibility	–	–
			Consensus through discussion

Across the Arctic SAR legal frameworks, this research notes the presence of many values, including human security, provisioning (making resources available) and responsibility (obligation and commitment to safeguard life). These values are repeated across the governance provisions for SAR, including documents such as the 1979 SAR Convention, the 2011 Arctic SAR Agreement and the 2017 Polar Code. In these documents, we find several values present including responsibility, efficiency and provisioning. In the more recent agreements, values connected to human security and environmental protection emerge. These values can be seen as a way to ensure that policies are aligned with desired outcomes and priorities and to provide a basis for monitoring and evaluation (Colebatch, 2006).

From search and rescue volunteers, both overlapping and a different set of value priorities emerge. In this group, we find values of connectivity, capability, human security provisioning responsiveness and wisdom. Notable for Arctic communities is that in this group, we find both Indigenous and non-Indigenous first responders. However, in their roles as SAR volunteers, they form a unique community and associated bonds—where belonging to a crew creates social bonds to facilitate rescue efforts (Aguirre et al., 1995). Given this, it seems appropriate that there is relative equal importance from all participants in these values. Many of these values are critical in their alignment with the core tasks of exercising and practising SAR.

At the intersection of resilience and values, we also consider the cultural context for the wider community in which the community of SAR volunteers is based. Drawing attention to the Inuit value system, which is of the highest relevance for the incidents discussed below, is another system to consider. This system, called "Inuit Qaujimaningit" which "refers to Inuit Traditional Knowledge as well as Inuit epistemology without reference to temporality", includes values summarised as respect and care for people (Inuuqatigiitsiarniq), openness and inclusivity (Tunnganarniq), serving the community (Pijitsirniq), consensus through discussion (Aajiiqatigiinniq), skill development through practice (Pilimmaksarniq), collaborative teamwork (Ikajuqtigiinniq), innovation and resourcefulness (Qanuqtuurniq), respect and care for the environment (Avatittinnik Kamatsiarniq), (NIRB, n.d.).

These value clusters show the essential character and priorities for conduct in the context of SAR, from three different dimensions: legal frameworks, organisational (albeit a volunteer community) and an Indigenous society. Although there is strong alignment across these sets, such as in the emphasis on ensuring safety and security, there are also examples of misalignment such as an emphasis on efficiency potentially undermining the potential for adaptive responses that can take either social or environmental priorities into account. As climate change erodes the stability and predictability of Arctic environmental conditions, it will be essential for SAR operational practices to also evolve—from both the legal and community approaches, while maintaining critical values essential to these communities of practice within their specific cultural contexts.

Stories from the SAR Landscape

While every coastal state has a responsibility under international law to provide for coordination of search and rescue, the implementation of this coordination is restricted by the assets (both human and technological) that underpin the response capacity. In Canada and elsewhere in the Arctic "Search and rescue (SAR) operations on land, water, and ice…are often complex and challenging due to austere environmental conditions, the strain they place on local resources, and the vast distances involved in responding" (Kikkert et al., 2023, p. 1). While most SAR systems have three levels of assets and responsibility within their planning, in the Canadian system, this tends to be divided into primary assets—which includes government resources (e.g. military), secondary assets, such as vessels of opportunity (e.g. commercial aeroplanes or AMVER vessels) and then finally, local volunteers (Canada Coast Guard, 2023). However, as most military assets are not readily available or located in the far north, frequently the initial response can quickly fall to local volunteers.

Yet, the volunteer layer of Arctic SAR has a great many challenges as it adapts to the new normal conditions for operations in the context of the changing climate. For example, "People who are willing to volunteer end up wearing a lot of different hats in the community. People are often willing to go out on rescues, but it is much harder to get them out for meetings, training, practice" (Kikkert & Lackenbauer, 2021, p. 48). In addition to all the activities that are required to maintain skill and coordination levels, it has been identified that there are also issues of inequality of resources among different types of volunteer organisations, the financial costs of participation (e.g. specialised clothing or equipment) that negatively impact less affluent volunteers, as well as problems with intergenerational transfer of geographic knowledge or mental health burdens from participating in SAR missions, especially those missions with negative outcomes (JUSTNORTH, 2023). Finally, local environmental knowledge and landscape reading skills are essential for SAR's success, however, the long-used signals and symbols are increasingly unreliable.

If you search for news stories of SAR in Arctic communities, you will find uplifting headlines of survival that read "hunter, survives 5-hour trek after falling through ice" and tell how the ice gave way and although well prepared with emergency supplies—these failed to work properly because of the cold (CBC, 2014). In the same article, however, you can also learn how weeks earlier in two separate incidents, others lost their lives. These types of incidents show that Indigenous knowledge, expertise and know-how are at a disjuncture with the Arctic environment as it is encountered today. With resilience including both psychological and functional features, these stories demonstrate both environmental know-how (i.e. knowledge of nature or survival skills) and environmental, and perhaps cultural, alienation.

We learn why this is through other articles that connect the incidents to climate change, such as, "Inuit crash through ice to death as warming bites", which posits that "global warming has made life more dangerous for Inuit hunters in the Arctic as they increasingly fall to their deaths through thinning ice sheets while pursuing seals and polar bears" (Reuters, 2007). Another article reminds us that the Inuit have thousands of years of experience and knowledge related to survival in the north and hunting on ice, but now "Inuit knowledge is vanishing with the ice" (Baraniuk, 2021). In a very short space of time, that knowledge has become less reliable, connecting to another grim "statistic" that in some places, every Inuit "knew someone who had died by falling through the ice and drowning" (Reuters, 2007)—possibly resulting in some form of "ecological grief" (Cunsolo & Ellis, 2018).

Through these stories and more, we can start to understand some of the big challenges between climate change, SAR and resilience in Arctic communities. The environmental state change that a warming Arctic brings to the landscape includes both one of danger and one of opportunity. Some of the dangers are illustrated in stories of misfortune above However, opportunities are highlighted in the growing number of cruise ships, fishing and other commercial activities in the Arctic—each one of these bringing potential for mass casualty incidents in Arctic communities. With search and rescue incidents having doubled (Clark et al., 2016) and maritime activities significantly increasing, the demand for SAR capacity at local levels has increased. And while capacity for provision has also improved in some locations (Kikkert & Lackenbauer, 2021)—the supply may not be equal to the demand.

Another challenge lies at the intersection of the allocation of legal responsibility, the use of ice as land and the instability brought through climate change. Although the stories draw attention to the potential to fall through ice—they do not illustrate well that ice serves as a different sort of territorial domain for the Inuit. This has been described as part of the Inuit homeland, as part of their mobility system and that "Inuit have lived on the land and ice-covered water in the Arctic and used the resources of the land and water to grow as a people. Inuit are deeply connected to not just the land but also the Arctic Ocean, as well as to everything in those environments" (Aporta & Watt, 2020, p. 188). This use is in stark contrast to SAR legal frameworks, which allocate responsibility for incident response in water to one authority and on land to another—which is particularly problematic when ice used as land fails to be as expected and a hunter falls into "water", that is impassable by a rescue boat because it is frozen. This evidence points to a disconnect between the relatively static regulatory frameworks and the operational environment, where the legal knowledge is blind to the environmental realities.

Resilience and the Landscape of Disjunction

It has been argued elsewhere that cultural values are the "bedrock of resilience" (Eggerman & Brick, 2010, p. 71) and are an important factor in understanding how individuals, communities and organisations respond to adversity. In particular, values are understood to "shape individual discrimination of experiences as either facilitative of growth or posing a barrier to personal development" (Ungar, 2012, p. 22). For communities in the Arctic with a strong set of values connected to identity, belonging to a group and in some cases, to their very existential survival, understanding values as a sort of compass can be critical in evaluating capacity for resilience.

With resilience understood as the ability of a complex system to maintain itself, despite disruptions and disturbances to some elements of the system, and in particular "to withstand external shocks" (Rana, 2020, p. 2). However, discontinuities can "reveal an aspect of resilience" (Allen et al., 2005, p. 961). This opens up the space to reflect on the discontinuities, or disjunctions that exist within this particular landscape: climate, values and Arctic communities to absorb these disjunctions, adapt to new realities and transform into a new form of equilibrium.

In the high latitudes of the planet that is called the Arctic, climate change is negatively affecting the major characteristic of the region: the cold temperatures that support ice, snow and permafrost. The durability of these features over a lengthy history has resulted in ecological life that thrives in this environment and in turn, human life that has an integrated relationship with this system. Given this, ice has been referred to as a "prominent life force" of the region and "there is no way to think of social and natural spaces as actually separate, both are formatted by the ice..." (Hastrup, 2021, p. 25). With the disappearance of the ice, the knowledge that has been developed to ensure survival in this environment is forced to adapt quickly and without the benefit of intergenerational wisdom.

The climate disjunction is not the only systemic shock experienced by Arctic communities. In the not-so-distant past, the entire existence of Arctic peoples was subject to transformative change in colonial policies and practices that shifted entire ways of knowing, being and doing across Indigenous communities in the North. This resulted in colonial structures that are both barriers and limits where "Indigenous communities have permanently lost land, mobility, and self-governance, making adaptation fundamentally more difficult" (Malik & Ford, 2025, p. 5). It has been recently argued that "Climate change—the newest instalment of developments produced elsewhere, with which Arctic people must struggle—is thereby just the newest wave of colonial violence and dispossession" (Krause, 2023, p. 170). However, with the climate crisis produced externally to Arctic communities, it is certainly a system shock requiring a response.

This is where the values of communities become key for resilience in the face of this disjunction which causes "erosion of traditional Inuit knowledge and land-based skills" (Ford et al., 2007, p. 150). Despite these circumstances, analysis shows that there is "significant adaptative capacity" in Arctic communities enabled by "a strong sense of collective community responsibility and mutual aid" (Ford et al., 2007, p. 150 and p. 154). However, government policies can impose limitations on this adaptative capacity through restrictions on hunting seasons, harvest quotas or even, labour laws restricting deviations from working hours in employed labour. These limitations can impose a contradiction been values and assorted needs, such as between practices that maintain cultural identity versus maintaining financial flows required for modern "conveniences" (e.g. electricity or housing loans).

Just as local values support adaptive capacity for knowledge systems that enable life in the Arctic, there is a similar connection to the values that underpin SAR regulatory systems, such as human security or responsibility, to ensure that are capable of ensuring resilience in Arctic communities. However, an inherent issue with regulatory frameworks connected to SAR is that they are not designed for the context of the Arctic environment—which also means that they are not adaptive to the circumstances that are found in the context of climate change. These frameworks, in their effort to create reliability, in fact impose a rigidity that does not align with the potential for local adaptability, which is vital for effective and efficient response to both the changing environmental state and the individual emergencies that emerge while communities learn to live with the changing climate.

Conclusion

When reflecting on resilience capacities, including SAR activities and value commitments, Arctic communities face deep challenges in the context of climate change. This chapter noted several key points of disjunction in a complex system of values, such as responsibility, capability or respect and care for the environment. A multi-faceted understanding of the complexity of this system and its underlying values is required to examine the adaptive capacity of Arctic communities at the intersection of SAR and climate change. Resilience in Arctic communities is premised on a system that is historically rich in meaning, flexible and responsive to the requirements and timescales of adaptation, and in particular—without compromising the core values that are fundamental to actors within the system.

To ensure that the legal frameworks and infrastructure that support SAR systems in the Arctic remain relevant in a changing Arctic, they must be able to align both with the values held by SAR volunteers and the communities that enable their efforts, as well as with the legal provisions underpinning SAR infrastructure. With the instability of the environment and pressures resulting from increasing economic activities in the region, these value systems may be up against pressures that are not able to withstand the rigidity of SAR regulations, with siloed responsibility and resource disparities. Without this alignment, the volunteer-driven SAR provision that emerges from Arctic communities may not be able to sufficiently adapt and without this capacity, limitations on its resilience are revealed. With SAR volunteers forming the backbone of Arctic SAR, any shortfalls in this domain and lack of response capacity to emergency events may ultimately lead to questions about the assumptions of responsibility held by Arctic states, and whether their sovereignty in the region is hollow because they cannot ensure safety and security of communities in the region.

References

Aguirre, B. E., Wenger, D., Glass, T. A., Diai-Murillo, M., & Vigo, G. (1995). The social organization of search and rescue: Evidence from the guadalajara gasoline explosion. *International Journal of Mass Emergencies & Disasters, 13*, 67–92. https://doi.org/10.1177/028072709501300105

Allen, C. R., Gunderson, L., & Johnson, A. R. (2005). The use of discontinuities and functional groups to assess relative resilience in complex systems. *Ecosystems, 8*, 958–966. https://doi.org/10.1007/s10021-005-0147-x

Arctic Council (2024). *Arctic shipping update: 37% increase in ships in the arctic over 10 years.* Available online at: https://arctic-council.org/news/increase-in-arctic-shipping/

Aporta, C., & Watt, C., 2020. Arctic waters as Inuit homeland, In T. Koivurova, E. G. Broderstad, D. Cambou, D. Dorough, & F. Stammler (Eds.), Routledge handbook of indigenous peoples in the arctic (pp. 187–205). Routledge. https://doi.org/10.4324/9780429270451-14

Baraniuk, C., 2021. *The Inuit knowledge vanishing with the ice*. BBC.

Barasa, E., Mbau, R., & Gilson, L. (2018). What is resilience and how can it be nurtured? A systematic review of empirical literature on organizational resilience. *Int J Health Policy Manag, 7*, 491–503. https://doi.org/10.15171/ijhpm.2018.06

Bohland, J. R., & Lawrence, J. (2020). Disaster resilience policy through the lens of culture and values. In J. M. Shultz, & A. Rechkemmer (Eds.), *The Oxford handbook of complex disaster risks and resilience*. Oxford University Press. https://doi.org/10.1093/oxfordhb/9780190466145.013.5

Canada Coast Guard. (2023). *Maritime search and rescue (SAR) in Canada [WWW Document]. Government of Canada*. https://www.ccg-gcc.gc.ca/publications/search-rescue-recherche-sauvetage/sar-canada-res-eng.html. Accessed September 21, 2024.

Cardona, O. D., van Aalst, M. K., Birkmann, J., & Forham, F. et al. (2012). *Determinants of risk: Exposure to vulnerability in managing the risks of extreme events and disasters to advance climate change adaptation, a special report of the working groups 1 and II of the Intergovernmental Panel on Climate Change (IPCC)*. Cambridge University Press.

Caymaz, E., Buyuksagnak, B.Y., & Ozsoy, B. (2023). Adaptive governance in integrating sustainability and resilience into the Arctic shipping routes. In *Supply chain operations in the Arctic* (pp. 279–300). Routledge. https://doi.org/10.4324/9781003218043-12

Chircop, A., Goerlandt, F., Aporta, C., & Pelot, R. (Eds.). (2020). *Governance of Arctic shipping: rethinking risk, human impacts and regulation*. Springer International Publishing.

Clark, D. G., & Ford, J. D. (2017). Emergency response in a rapidly changing Arctic. *CMAJ, 189*, E135–E136. https://doi.org/10.1503/cmaj.161085

Clark, D. G., Ford, J. D., Berrang-Ford, L., Pearce, T., Kowal, S., & Gough, W. A. (2016). The role of environmental factors in search and rescue incidents in Nunavut, Canada. *Public Health, 137*, 44–49. https://doi.org/10.1016/j.puhe.2016.06.003

Colebatch, H. K. (2006). *Beyond the policy cycle: The policy process in Australia*. Allen & Unwin.

Cuaton, G. P., & Su, Y. (2020). Local-indigenous knowledge on disaster risk reduction: Insights from the Mamanwa indigenous peoples in Basey, Samar after Typhoon Haiyan in the Philippines. *International Journal of Disaster Risk Reduction, 48*, Article 101596. https://doi.org/10.1016/j.ijdrr.2020.101596

Cunsolo, A., & Ellis, N. R. (2018). Ecological grief as a mental health response to climate change-related loss. *Nature Clim Change, 8*, 275–281. https://doi.org/10.1038/s41558-018-0092-2

Daskon, C. D. (2010). Cultural resilience—The roles of cultural traditions in sustaining rural livelihoods: A case study from rural Kandyan villages in Central Sri Lanka. *Sustainability, 2*, 1080–1100. https://doi.org/10.3390/su2041080

Drzewiecki, D. M., Wavering, H. M., Milbrath, G. R., Freeman, V. L., & Lin, J. Y. (2020). The association between educational attainment and resilience to natural hazard-induced disasters in the West Indies: St. Kitts & Nevis. *International Journal of Disaster Risk Reduction, 47*, 101637. https://doi.org/10.1016/j.ijdrr.2020.101637

Eggerman, M., & Panter-Brick, C. (2010). Suffering, hope, and entrapment: Resilience and cultural values in Afghanistan. *Social Science & Medicine, 71*, 71–83. https://doi.org/10.1016/j.socscimed.2010.03.023

Ford, J., Pearce, T., Smit, B., Wandel, J., Allurut, M., Shappa, K., Ittusujurat, H., & Qrunnut, K. (2007). Reducing vulnerability to climate change in the arctic: The case of nunavut, Canada. *ARCTIC 60*, 150–166. https://doi.org/10.14430/arctic240

Graybill, J. K., & Petrov, A. N. (Eds.). (2020). *Arctic sustainability, key methodologies and knowledge domains: A synthesis of knowledge I* (1st edn.). Routledge | Series: Routledge research in Polar Regions. https://doi.org/10.4324/9780429277016

Hastrup, K. (2021). Living (with) Ice: Geo-sociality in the high arctic. *J North Stud, 14*, 25–41. https://doi.org/10.36368/jns.v14i2.984

JUSTNORTH. (2023). *Project brief: Northern Seas, global connections: Shipping, search & rescue and small communities in Canada & Norway.*

Kekes, J. (1993). *The morality of pluralism.* 4. print., and 1. paperback print. ed, Princeton paperbacks. Princeton Univ. Press.

Kikkert, P., & Lackenbauer, P. W. (2021). Search and rescue, climate change, and the expansion of the coast guard auxiliary in Inuit Nunangat / the Canadian Arctic. *CJEM, 1*. https://doi.org/10.25071/vaa86009

Kikkert, Peter, Quigley, John and Belton, Ian and Brown, Robert and Ennis, Kerri-Ann and Goerlandt, Floris and Howick, Susan and Lackenbauer, P. Whitney and Mak, Lawrence and Pedersen, Calvin and Pelot, Ronald and Shan, Desai and Walls, Lesley and Wright, George (2023). *Addressing the challenges to search and rescue operations caused by ice conditions in Nunavut, Canada, in: 27th International Conference on Port and Ocean Engineering under Arctic Conditions.* Presented at the 27th International Conference on

Port and Ocean Engineering under Arctic Conditions (pp. 1–16). University of Strathclyde.

Koliou, M., Van De Lindt, J. W., McAllister, T. P., Ellingwood, B. R., Dillard, M., & Cutler, H. (2018). State of the research in community resilience: Progress and challenges. *Sustainable and Resilient Infrastructure, 5*, 131–151. https://doi.org/10.1080/23789689.2017.1418547

Krause, F. (2023). Climate change as colonial echo in the Canadian Arctic. In *Anthropology and climate change* (pp. 170–180). Routledge. https://doi.org/10.4324/9781003242499-14

Malik, I. H., & Ford, J. D. (2025). Barriers and limits to adaptation in the Arctic. *Current Opinion in Environmental Sustainability, 73*, Article 101519. https://doi.org/10.1016/j.cosust.2025.101519

Malik, I. H., Ford, J. D., Clark, D. G., Pearce, T. (2024). Climate change, mass casualty incidents, and emergency response in the Arctic. *Environ. Res.: Infrastruct. Sustain., 4*, 043002. https://doi.org/10.1088/2634-4505/ad8fcc

Mayar, K., Carmichael, D. G., & Shen, X. (2022). Resilience and systems—A review. *Sustainability, 14*, 8327. https://doi.org/10.3390/su14148327

McMillen, H.L., Ticktin, T., Friedlander, A., Jupiter, S.D., Thaman, R., Campbell, J., Veitayaki, J., Giambelluca, T., Nihmei, S., Rupeni, E., Apis-Overhoff, L., Aalbersberg, W., Orcherton, D.F. (2014). Small islands, valuable insights: systems of customary resource use and resilience to climate change in the Pacific. *E&S, 19*, art44. https://doi.org/10.5751/ES-06937-190444

NIRB (n.d.). *Inuit Qaujimajatuqangit* [WWW Document]. Nunavut Impact Review Board. https://www.nirb.ca/inuit-qaujimajatuqangit. Accessed September 20, 2024.

O'Brien, K., Pelling, M., Patwardhan, A., Hallegatte, S., Maskrey, A., Oki, T., Oswold-Spring, Ú., Wilbanks, T., Yanda, P.Z., 2012. Toward a sustainable and resilient future. In B. F. Christopher, V. Barros, T. F. Stocker, & Q. Dahe, Q. (Eds.), *Managing the Risks of extreme events and disasters to advance climate change adaptation: Special report of the intergovernmental panel on climate change*. Cambridge University Press. https://doi.org/10.1017/CBO9781139177245

Posch, E., Höferl, K. M., Steiger, R., Bell, R., & Gurung, L. (2019). Ke garne? How values and worldviews influence resilience to natural hazards: A case study from mustang, Nepal. *Mountain Research and Development, 39*,. https://doi.org/10.1659/MRD-JOURNAL-D-19-00005.1

Rana, I. A. (2020). Disaster and climate change resilience: A bibliometric analysis. *International Journal of Disaster Risk Reduction, 50*, Article 101839. https://doi.org/10.1016/j.ijdrr.2020.101839

Rantanen, M., A., Yu Karpechko, Lipponen, A., Nordling, K., Hyvärinen, O., Ruosteenoja, K., Vihma, T., & Laaksonen, A. (2022). The Arctic has warmed nearly four times faster than the globe since 1979. *Commun Earth Environ, 3*, 168. https://doi.org/10.1038/s43247-022-00498-3

Reuters (2007). *Inuit crash through ice to death as warming bites*. Reuters.

Rokeach, M. (1973). *The nature of human values*. Free Press.

Schwartz, S.H., 1992. Universals in the content and structure of values: Theoretical advances and empirical tests in 20 countries. In *Advances in experimental social psychology* (pp. 1–65). Elsevier. https://doi.org/10.1016/S0065-2601(08)60281-6

Skura, E. (2014). *Laimiki Pewatualuk, Nunavut hunter, survives 5-hour trek after falling through ice*. CBC-Canada Broadcasting Corporation.

Stark, A. (2014). Bureaucrative values and resilience: An exploration of crisis management adaptation. *Public Administration, 92*, 692–706. https://doi.org/10.1111/padm.12085

Ungar, M. (2012). Social ecologies and their contribution to resilience. In M. Ungar (Ed.), *The social ecology of resilience* (pp. 13–31). Springer. https://doi.org/10.1007/978-1-4614-0586-3_2

Wood-Donnelly, C. (2013). The arctic search and rescue agreement: Text, framing and logics. *The Yearbook of Polar Law Online, 5*, 299–318. https://doi.org/10.1163/22116427-91000127

Wood-Donnelly, C. (2019). *Performing Arctic sovereignty: Policy and visual narratives* (S.l). Routledge.

Wood-Donnelly, C. (2024). *Negotiating the Arctic: Sustainability, governance and environmental justice.* The Fletcher Forum of World Affairs 48.

13

Interrogating Resilience: A Critical Analysis of its Use and Impact on Indigenous Discourses in Australia and the Arctic

Sakshi and Darren McCauley

Introduction

'Resilience' is a term and concept with a multidisciplinary presence. Whilst having a shared vocabulary of what it entails, the concept has varying connotations in physical sciences, human geography, psychology, and ecology, amongst others (Bahadur et al., 2013; Jolly, 2020). It has also made recurring appearances in the climate change literature. The term has also become an organising force for climate action, imaginaries of climate justice, and, in some spaces, an integral part of describing climate futures. The term has been found to have robust application and utility in socio-economic responsiveness to disaster risk reduction and climate adaptation planning. Community involvement and inclusion of local knowledge in various spheres of governance also demand

Sakshi (✉) · D. McCauley
Newcastle Law School, Newcastle University, Newcastle Upon Tyne, UK
e-mail: Sakshi@newcastle.ac.uk

T. Skillington and A. Setti (eds.), *Climate Change Resilience Across Societal Contexts*, Energy, Climate and the Environment,
https://doi.org/10.1007/978-3-031-95020-9_13

a detailed understanding of resilience as a concept and analytical framework. However, 'resilience' plays a deeper political role in where it is used and where it is ignored. Whilst 'resilience' has its utilities, the socio-political and juridical contexts in which the term is deployed tend to undermine the necessary actions demanded by the situations and resistances offered by individuals and stakeholders expected to be 'resilient', such as the Indigenous peoples (Jackson, 2024).

In this paper, we explore two iterations of the concept of resilience. First, we examine the use of resilience in Australia to argue how the concept distracts from the continuing generation of stress factors that make resilience necessary in the first place. The case study indicates the lack of understanding of historical inequalities pertaining to Indigenous peoples that makes resilience impossible or harder to deploy as frequently as it is demanded. We explore the state-endorsed fossil fuel drive in the Northern Territory, Australia and the Indigenous resistance to land expropriation has denuded resilience of its usefulness. Second, we examine resilience as a useful concept deliberately underserved in the Arctic discourse. To this end, we advance arguments that illustrate how socio-ecological and relational articulation of resilience through Indigenous voices are undermined by either complete erasure in Western knowledge production or subject to strategic limiting in commercial ventures. The case study focuses on the Polar Code as a space for many forms of epistemic and material contestations.

Climate Crises and Responses in Australia

Since 2020, Australia has experienced an exponential increase in extreme weather events. The Black Summer bushfires of 2020–21, catastrophic flooding in Eastern Australia due to three consecutive La Niñas (2022–23), and the wide-ranging impact of Cyclone Jasper in Queensland, amongst other significant climatic events, had profound social, economic, and psychological impacts on people across the board (Gillett et al., 2023). Whilst the extreme weather mobilised people to demand immediate responses in terms of relief and emergency support, they also became an embodiment of ongoing climate crises and opened avenues for broader conversation around Australia's climate action plan. The collapse of the ice shelf and low levels of Antarctic sea ice became as important as local communities suffering from floods in Lismore or extreme heatwaves in the Pilbara and Northern Territory. The immediate response of the Commonwealth and State governments was to commission reports and create Royal Commissions that looked at resilience and disaster preparedness in Australia whilst pushing for the 'build back better' narrative (Australian Institute for Disaster Resilience, 2023). There was also a sudden resurgence of interest in Indigenous leadership in the conservation and cultural management of land and waters (Mosurska et al., 2023).

Admittedly, some useful findings from the royal commissions and select committees helped determine the courses of action in frequent extreme weather events. 'The Royal Commission into National Natural Disaster Arrangements (2020)' and the Select Committee report 'Boots on the Ground: Raising Resilience' (2024) exemplify how disasters created opportunities for reactive knowledge production. However, as Taylor and Crawford (2024) show, these reports shift the burden on people, volunteers, and, at best, planning mechanisms insofar as having an inclusive response and recovery workforce. For instance, the Special Committee suggested formalising the inclusion and training of community organisations that had played a vital role in the previous disasters (Select Committee, 2024). The processes of official knowledge production, however (seemingly) apolitical, reinforce the idea that disasters are singularly environmental.

Consequently, mainstream governmentality leads to misleading perceptions that environmental or climatic disasters are not socially and politically generated (Tierney, 2018). Australia's environmental governance has unequivocally and brazenly taken a neoliberal turn by relying on resilience and preparedness as the only two frameworks to determine public behaviour. Whilst settler colonialism and extractive capitalism have always been the dominant features of the Australian body politic, 'resilience' as an anchoring concept in environmental and climate policy evades scrutiny from pre-existing inequalities and aggravated forms of environmental injustices triggered by extreme events or climate change, more broadly (Cappelli et al., 2021).

Extracting Resilience, Erasing Self-Determination

Due to the mainstream prevalence of scientific knowledge regarding climate change and climate change-induced extreme weather events, most public opinions are now at odds with the political willingness towards climate action. The rifts and fractures of environmental governmentality in Australia, primarily shaped by disaster risk reduction and response, became more concrete due to increasing resentment against the broad-brushed endorsement of fossil fuels. Since the current Labour government came into power, Australia has announced seven new coal mines and various oil exploration permits, especially in the ecologically fragile and culturally sensitive Indigenous lands across Western Australia and Northern Territory (Climate Council, 2024). It is of interest to the arguments here that Indigenous peoples have been the common denominators in the responses to extreme weather events and the relentless expansion of extractive activities. Whilst Indigenous knowledge holders are deemed resilient, there is an organised silence around the dispossession of Indigenous peoples and the systematic devaluation of Indigenous ways of life within ongoing settler colonialism in Australia (Watson, 2012). Although Indigenous people are resilient in surviving settler colonialism and genocide, at present, they have been subjected to a variety of stress factors that make resilience a less useful and viable concept. First Nations across Australia are continually battling the degradation of traditional lands as well as formal land rights.

Despite the tokenistic nod to Indigenous leadership in land management and environmental conservation, the increased extractive activities have resulted in First Nations offering resistance through political and legal actions. These actions by Indigenous communities, especially strategic litigation, have increased in the last five years. Various forms of environmental litigation led by First Nations, even if strategic, come at an enormous economic and political cost. Especially since the legal adversaries in Indigenous litigation tend to be either the state or the massive oil and gas corporations, such as Woodside, Fortescue, BHP, and Rio Tinto, the costs—economic or otherwise—fall disproportionately on the Indigenous litigants. Despite the resources and time that goes into litigation, the juridical spaces do not always guarantee success, and the First Nations are left with the loss of cultural heritage and spaces to reassert their sovereignty. An instance of such failure may be seen in the destruction of sacred sites such as Jukkan Gorge and the repetition of similar events in Burrup and Flinders Range.[1] In addition, Indigenous communities are also expected to establish their land and spiritual rights by constantly contesting their claims before courts and often find themselves opposed by both the state and the extractive industries. One such instance is the recent litigation against oil giant Santos, resulting in serious hindrances to Indigenous environmental justice and resilience.

Santos Limited was established in 1954 for oil exploration in South Australia and Northern Territory. Whilst Santos is no different from any other energy corporation that constitutes the infrastructures of extraction and accumulation in Australia, some operations of Santos have attracted judicial attention. In particular, Santos' Barossa project in the Northern Territory attracted staunch opposition from the Tiwi Islanders. In September 2022, in Tipakalippa v National Offshore Petroleum Safety and Environmental Management Authority (No 2) ('*Tipakalippa*'),[2] the Federal Court of Australia set aside National Offshore Petroleum Safety and Environmental Management Authority's (NOPSEMA) decision to accept the mining giant Santos' proposed Barossa project, a $5.2 billion offshore gas development project, set to run between 2022 and 2025. The project will likely emit 380 million tonnes of greenhouse gas pollution over its lifetime, equivalent to 81% of Australia's total emissions in 2022 (Verstegen & Campbell, 2023). Bromberg J, known for his climate-friendly decisions, handed out the decision. The *Tipakalippa* judgement was decided on the grounds that there was no adequate consultation with the First Nations and that the sea country material—Indigenous knowledge pertaining to the cultural and environmental significance of the sea—was disregarded.

Following the decision, Santos revised its plan. In December 2023, NOPSEMA accepted the revised drilling plan proposed by Santos. By its own assessment, Santos concluded that they had carried out 'further extensive consultation with Tiwi Island people and other relevant persons consistent with the applicable regulations'. Simon Munkara, a Tiwi Islander, challenged this revised plan and NOPSEMA's decision to accept the same, resulting in *Munkara v Santos NA Barossa Pty Ltd (No.3)* ('Munkara').[3] Whilst the issues here were similar to the claims in *Tipakalippa*, the claims in *Munkara* made it categorically appear as a cultural heritage case spearheading environmental issues. Here, the applicants (Simon Munkara along with members of Munupi and Malawu clan as second and third applicants) alleged that the proposed plan for construction of the pipelines for the Barossa Project resulted in 'the occurrence of a significant new environmental impact of risk' under regulation 17(6) of the Offshore Petroleum and Greenhouse Gas Storage (Environment) Regulations 2009 (Cth). It was also alleged that there were several significant cultural features around the pipeline route. They contended that the spiritual connection that the Jikilaruwu, Munupi, and Malawu people have to the area of sea country through which the pipeline will pass was significant and that activities of Santos had resulted in many forms of damage to those relationships, thereby alluding to the ancestral catastrophes that do not necessarily have legal recourses. The claimants relied on the travel of an ancestral being, a rainbow serpent known as Ampiji. The applicants alleged that Ampiji is the caretaker of the sea and that the proposed pipeline will not only disturb her during her travels but also cause her to create calamities that may harm the Tiwi people. In addition, the applicants also stated that the song line of significance to the Jikilaruwu people—a shape-shifting ancestral known as Jirakupai or the Crocodile Man–was in the vicinity of the pipeline. Therefore, the proposed activities were claimed to disrupt and even anger the Crocodile Man.

Unfortunately, amongst other grounds, the Court relied on the disagreement between two groups of Tiwi Islanders to conclude that the claim must be dismissed. Santos has resumed operations since then, and the rich cultural heritage and environmental integrity of the region are under threat of permanent and irreversible damage. The economic philosophy of prioritising economic growth at any cost and absolute disregard for Indigenous knowledge where it stands in the way of extractivism contradicts the celebration of Indigenous resilience only where it is convenient. Indigenous stewardship in biodiversity conservation or fire management cannot exist in isolation whilst the conditions that aggravate climate change-induced extreme weather events or encourage fossil fuel production continue to decimate Indigenous land and knowledge forms (Williamson, 2021).

Whilst the Santos case may be just another instance of unsuccessful litigation, Australia's recent juridical and political history teems with examples of Indigenous communities taking on fossil and extractive industries to fight for social and climate justice for all. These endeavours strain and often destroy the fabric of resilience even where neoliberal governmentality maintains the façade that all resilience is indestructible and desirable (Jackson, 2024). Indigenous peoples' resistance is a robust example of why resistance to extractivism and the struggle for justice is more important than undefined, uncritical deference to resilience as a vehicle for environmental justice. The following case study from the Arctic reinforces the call for a more profound understanding of resilience.

Resilience, Indigenous Discourses, and the Arctic

In Western science-driven climate narratives, symbols of resilience are notably absent from Arctic discourse. This narrative often neglects Indigenous perspectives, creating a rift between global climate science discourse and local realities. Indigenous voices provide insights into the socio-ecological relationships that are crucial for a comprehensive understanding of resilience in the Arctic. These perspectives are frequently overshadowed by Western narratives that prioritise global conservation efforts over local needs (Tam et al., 2021). This marginalisation illustrates, above all, a tension between different understandings of resilience.

Resilience in the Arctic is a layered concept shaped by the area's distinct vulnerabilities and varied geopolitical landscapes. The region's environmental fragility and cultural diversity demand a nuanced understanding of resilience. In the Canadian Arctic, for instance, resilience is pursued through initiatives like the Low-Impact Shipping Corridors. These corridors are developed through collaboration with Inuit communities, highlighting a socio-ecological approach to resilience. Input from these communities is meant to ensure that shipping routes are adjusted to avoid disrupting marine ecosystems and Indigenous livelihoods.

Measures such as seasonal restrictions are recommended to safeguard marine mammals during crucial periods (Dawson et al., 2020). This approach shows one isolated example, that we return to below, of incorporating traditional knowledge into modern strategies to protect both the environment and cultural practices.

Contrastingly, the Republic of Sakha in Russia demonstrates resilience through social networks and traditions. Indigenous communities there have cultivated robust sharing systems that enable adaptation to climate change. These networks allow for the mobilisation of resources amidst economic and environmental pressures, pointing to the importance of social cohesion. Adaptation involves integrating modern technologies like social media to strengthen these networks. This adaptation reflects the importance of local knowledge and community-led initiatives in fostering resilience (Solovyeva & Kuklina, 2020). Such practices emphasise resilience through community bonds over technological reliance, highlighting a different aspect of resilience compared to Western interpretations.

Resilience in the Arctic is not just a response to environmental challenges. It also involves socio-economic considerations and respect for traditional knowledge. These examples reveal the contested nature of resilience in the Arctic, reflecting a spectrum of interpretations that range from commercial development to socio-ecological sustainability. We explore these contestations in relation to Arctic shipping specifically as a practical example, with a continued focus on Canada and Russia with other examples.

The Example of Arctic Shipping: A Commercial-Only Interpretation of Resilience?

The governance of shipping in the Arctic exemplifies the discursive contestation between commercial interests and Indigenous understandings of resilience. In Canada, there is a notable shift towards a more inclusive governance framework that recognises the importance of

Indigenous participation in maritime issues. This reflects an acknowledgement of Indigenous self-governance rights and the benefits of their involvement in decision-making processes. The Northern Low-Impact Shipping Corridors initiative, introduced above, is a case in point. Expanding further, it embraces the need for Inuit involvement in governance and highlights environmental protection and emergency response. Through their participation, the initiative embodies a reported increased appreciation for Indigenous perspectives that is integral to fostering resilience in the Arctic (Wang & Aporta, 2024).

In contrast, the Russian Northern Sea Route (NSR) emphasises economic opportunities derived from increased shipping. The route underscores the rising cargo volumes that indicate its commercial importance. However, this expansion often places Indigenous communities in precarious socio-economic positions as their traditional livelihoods are under threat. The growth of commercial shipping presents challenges that necessitate careful balancing with environmental and cultural considerations, which are found to be ignored (Huntington et al., 2023; Konnov et al., 2022). Without such a balance, local Indigenous resilience is compromised, as the development outpaces the mechanisms that protect Indigenous rights and livelihoods.

The distinction between Canada's collaborative governance approach and Russia's economically focused model underscores divergent paths in Arctic shipping governance. The Canadian approach illustrates how the integration of Indigenous rights into governance frameworks supports, in some way, and not without controversy, resilience by involving community knowledge and perspectives. This success is also comparative to Russia. In absolute terms, Canada has a mixed record in undertaking a collaborative governance approach as evidenced in von der Porten and de Loë (2013). Both examples from Canada and Russia reveal the necessity for international cooperation to navigate these complexities effectively. A precautionary approach is needed to mitigate adverse effects on the environment and Indigenous cultures. This is sadly lacking in Arctic contexts. We focus on one such mechanism, namely the Polar Code.

The Application of the Polar Code

The Polar Code, under the guidance of the International Maritime Organization (IMO), exemplifies the intersection of commercial and Indigenous interpretations of resilience in Arctic shipping governance. Its implementation across Arctic waters addresses the imperative for maritime safety and environmental protection, yet it simultaneously reveals distinct priorities between commercial interests and Indigenous communities.

Canada's adoption of the Arctic Shipping Safety and Pollution Prevention Regulations (ASSPPR) reflects an intent to balance global regulatory practices with national interests. Incorporating the Polar Code, these regulations aim to harmonise international commitments with Canada's unilateral coastal authority. A fundamental aspect of this is their emphasis on safeguarding the Arctic environment, aligning with Indigenous priorities of environmental stewardship and sustainable development. The ASSPPR's focus on protecting ecologically sensitive areas demonstrates Canada's commitment to long-term resilience, which resonates with Indigenous communities that view resilience through environmental and cultural lenses (Bartenstein, 2019). In contrast, Russia's contribution to the development of the Polar Code brings to light its focus on sovereignty and dominance over the Northern Sea Route. This approach emphasises strict navigation protocols and an authoritative decision-making framework, underlining a perception of resilience as maintaining control over strategic resources and routes. By instituting stringent safety measures, Russia seeks to ensure resilience against the harsh and unpredictable Arctic conditions whilst pursuing economic and geopolitical goals (Solski, 2023). This approach lacks any real engagement with indigenous narratives.

The geographical application of the Polar Code warrants consideration. Currently, its designated areas do not fully encompass the dynamic Arctic ecosystem, such as the Bering Sea and Sea of Okhotsk. The retreating ice coverage in these waters increases the likelihood of marine accidents, raising questions about the Code's adequacy in ensuring maritime resilience against climatic and ecological shifts. An adaptation of its parameters is crucial to mitigate risks and uphold safety standards across all vulnerable regions (Karahalil et al., 2020). From a commercial standpoint, the Polar Code is perceived as a vital benchmark for operational safety and conserving marine ecosystems, particularly by enterprises operating within the Arctic. Yet this view is tempered by concerns about the Code's restrictive scope and the economic implications of stringent compliance requirements. Commercial entities often stress the operational difficulties and financial burdens associated with meeting the Code's regulations, prioritising cost-efficiency and streamlining of Arctic shipping practices (Pic et al., 2021).

Indigenous perspectives regard the code as an instrument that should be designed to preserve cultural integrity and prevent environmental degradation. This juxtaposition highlights the nuanced interpretations of resilience where commercial bodies prioritise economic viability, whereas Indigenous communities emphasise ecological preservation and cultural continuity. These diverse perspectives necessitate a new continuous and adaptive dialogue within the Polar Code framework, ensuring it evolves to recognise and reconcile the varied resilience objectives of its stakeholders. The Polar Code serves as a critical point of reference in Arctic maritime governance. Its comprehensive application requires continual refinement to balance the dichotomy between commercial advancement and Indigenous cultural preservation. Ongoing adaptation is needed to ensure the framework meets the complex demands of Arctic navigation whilst promoting an inclusive resilience narrative that respects the diverse priorities of community and commercial actors alike.

Re-Imagining Resilience in Arctic Shipping

The discourse on resilience in Arctic shipping is significantly dominated by commercial interests. This bias prioritises economic advantages such as lower transportation costs and improved trade routes, often at the expense of Indigenous perspectives (Tiller et al., 2022). The focus on economic benefits overshadows the potential negative impacts on Indigenous communities, including disruptions to traditional hunting practices and cultural activities. Such an approach fails to address the broader socio-ecological challenges faced by these communities. Therefore, a shift towards more inclusive governance processes that respect and integrate Indigenous knowledge is essential. Digital technologies are becoming a powerful tool for Inuit organisations to challenge narratives dominated by Western perspectives on resilience and adaptation. Utilising online platforms, Inuit communities are increasingly vocal about their cultural and environmental values, advocating for a reframing of resilience to better reflect their lived experiences and priorities (Youngblood, 2019). Earthjustice is a leading example where such platforms have been established to focus on providing voice to resilience on health-related matters, as outlined in Youngblood (2019). This highlights the critical need for governance models that not only incorporate Indigenous viewpoints but also focus on achieving just outcomes. These outcomes should go beyond mere procedural fairness to ensuring the substantive rights of Indigenous peoples are respected and upheld.

A participatory scenarios workshop with Alaska Indigenous youth sheds light on the potential for creating inclusive decision-making frameworks. These young individuals express a deep-rooted connection to their land and communities. Their involvement in governance can drive future frameworks that factor in economic concerns whilst remaining deeply committed to cultural continuity and ecological preservation (Cost & Lovecraft, 2021). Engaging Indigenous youth in governance discussions paves the way for sustainable practices that align with both community values and long-term environmental health. The case of Arctic shipping exemplifies the urgency for governance structures that prioritise Indigenous knowledge systems. It demonstrates the necessity of outcomes that are both environmentally sound and culturally respectful. Legal and governance processes need to be reframed to ensure that they reflect the core values and aspirations of Arctic Indigenous communities. This will require meaningful engagement, and new ways of thinking.

Conclusion

The exploration of resilience within Indigenous discourses in Australia and the Arctic uncovers significant tensions between its theoretical applications and practical realities. The concept of resilience, though initially intended as a framework for adaptation and survival, often becomes a diversion from addressing the root causes of stress and inequality that Indigenous communities face. In Australia, the notion of resilience is frequently invoked in response to climate crises and natural disasters. However, this focus obscures the systemic issues of land dispossession and cultural erosion faced by Indigenous peoples. The state-driven emphasis on fossil fuel expansion further complicates this narrative, as it directly contradicts Indigenous efforts to protect their lands and assert their rights. Legal challenges, such as those involving Santos, illustrate the profound economic and cultural costs borne by Indigenous communities in their pursuit of justice and environmental integrity. These cases highlight the limitations of traditional resilience paradigms, which inadequately address the deeper socio-political dynamics at play.

In the Arctic, resilience is similarly contested, with Indigenous perspectives often marginalised within broader climate and environmental discussions. Western narratives tend to prioritise global conservation objectives over local needs, thereby overshadowing the rich socio-ecological insights offered by Indigenous communities. The examination of Arctic shipping governance exemplifies these tensions, where commercial interests frequently eclipse Indigenous rights and knowledge systems. Whilst Canada demonstrates some progress towards inclusive governance by incorporating Indigenous participation in maritime issues, the Russian approach shows a prevailing focus on economic gains. This disparity reveals the challenge of balancing commercial development with the preservation of Indigenous cultural and environmental priorities.

Re-imagining resilience requires moving beyond a one-size-fits-all approach. It demands an appreciation of the diverse interpretations and applications of resilience as experienced by Indigenous peoples across different geographies. By acknowledging the unique challenges and contributions of Indigenous communities, policymakers and stakeholders can develop more nuanced and responsive strategies. In doing so, resilience can be transformed from a superficial concept into a meaningful tool for empowerment and justice, aligned with the values and realities of those it is meant to serve. This redefined understanding of resilience has the potential to bridge the gap between global environmental imperatives and the lived experiences of Indigenous peoples.

Notes

1. In 2020, during the expansion of its iron ore mining operations, Rio Tinto destroyed a sacred site belonging to Puutu Kunti Kurrama and Pinikura people. Although Rio Tinto was fully aware of the cultural and spiritual significance of the site, it was not held liable as the expansion operations were legally permitted. Subsequently, although heritage laws were strengthened across Australia, they fail to check activities that are considered economically significant.
2. Tipakalippa v National Offshore Petroleum Safety and Environmental Management Authority (No 2) [2022] FCA 1121.
3. Munkara v Santos NA Barossa Pty Ltd (No.3),[2024] FCA 9.

References

Australian Institute for Disaster Resilience (2023). *Australia's riskscape: A companion to the major incidents report 2022–23 and the systemic disaster risk handbook.*

Bahadur, A. V., Ibrahim, M., & Tanner, T. (2013). Characterising resilience: Unpacking the concept for tackling climate change and development. *Climate and Development, 5*(1), 55–65.

Bartenstein, K. (2019). Between the polar code and article 234: The balance in Canada's Arctic shipping safety and pollution prevention regulations. *Ocean Development and International Law, 50*, 335–362. https://doi.org/10.1080/00908320.2019.1617932

Cappelli, F., Valeria, C., Consoli, D. (2021). The trap of climate change-induced "natural" disasters and inequality. *Global Environmental Change, 70*, 102329.

Climate Council. (2024). *Our kids will not forgive coal mine approvals*. https://www.climatecouncil.org.au/resources/our-kids-will-not-forgive-coal-mine-approval/

Cost, D., & Lovecraft, A. (2021). Scenarios development with Alaska's Arctic Indigenous youth: Perceptions of healthy sustainable futures in the Northwest Arctic Borough. *Polar Geography, 44*, 112–135. https://doi.org/10.1080/1088937X.2020.1755906

Dawson, J., Carter, N., van Luijk, N., Parker, C., Weber, M., Cook, A., Grey, K., & Provencher, J. (2020). Infusing Inuit and local knowledge into the low impact shipping corridors: An adaptation to increased shipping activity and climate change in Arctic Canada. *Environmental Science and Policy, 105*, 19–36. https://doi.org/10.1016/j.envsci.2019.11.013

Gillett, Z., Shao, Y., Parker, T., Barnes, M., Reid, K., Arblaster, J., Meyer, A., Pitman, A., Yang, D., Raupach, T., Brown, A., Hitchcock, S., & El Rafei, M. (2023). *The state of weather and climate extremes-2022. ARC centre of excellence for climate extremes*. UNSW.

Huntington, H., Olsen, J., Zdor, E., Zagorskiy, A., Shin, H., Romanenko, O., Kaltenborn, B., Dawson, J., Davies, J., & Abou-Abbsi, E. (2023). Effects of Arctic commercial shipping on environments and communities: context, governance, priorities. *Transportation Research Part D-Transport and Environment, 118*, Article 10373. https://doi.org/10.1016/j.trd.2023.103731

Jackson, G. (2024). The influence of climate resilience governmentality on vulnerability in regional Australia. *Environment and Planning e: Nature and Space, 7*(3), 1098–1121.

Jolly, M. (2020). Bushfires, supercyclones and 'resilience': Is it being weaponised to deflect blame in our climate crisis? *Scottish Geographical Journal, 136*(1–4), 81–90.

Karahalil, M., Ozsoy, B., & Oktar, O. (2020). Polar Code application areas in the Arctic. *WMU Journal of Maritime Affairs, 19*, 219–234. https://doi.org/10.1007/s13437-020-00200-4

Konnov, A., Khmelnitskaya, Y., Dugina, M., Borzenko, T., & Tysiachniouk, M. (2022). Traditional livelihood, unstable environment: adaptation of traditional fishing and reindeer herding to environmental change in the Russian arctic. *Sustainability, 14*, Article 12640. https://doi.org/10.3390/su141912640

Mosurska, A., Clark-Ginsberg, A., Sallu, S., & Ford, J. (2023). Disasters and indigenous peoples: A critical discourse analysis of the expert news media. *Environment and Planning e: Nature and Space, 6*(1), 178–201.

Pic, P., Babin, J., Lasserre, F., Huang, L., & Bartenstein, K. (2021). The Polar code and Canada's regulations on Arctic navigation: Shipping companies' perceptions of the new legal environment. *Polar Journal, 11*, 95–117. https://doi.org/10.1080/2154896X.2021.1889838

Select Committee on Australia's Disaster Resilience. (2024). *Boots on the ground: Raising resilience*. Commonwealth of Australia.

Solovyeva, V., & Kuklina, V. (2020). Resilience in a changing world: Indigenous sharing networks in the Republic of Sakha (Yakutia*). Polar Record, 56*, Article PII S0032247420000406. https://doi.org/10.1017/S0032247420000406

Solski, J. (2023). The polar code process and sovereignty bargains: comparing the approaches of Canada and Russia to POLARIS. *Ocean Development and International Law, 54*, 111–134. https://doi.org/10.1080/00908320.2023.2190940

Tam, C., Chew, S., Carvalho, A., & Doyle, J. (2021). Climate change totems and discursive hegemony over the Arctic. *Frontiers in Communication, 6*, Article 518759. https://doi.org/10.3389/fcomm.2021.518759

Taylor, M., & Crawford, F. (2024, August 9). Disaster season looms, but the senate inquiry has failed to empower communities. *The Conversation.* Available at: https://theconversation.com/disaster-season-looms-but-the-senate-inquiry-has-failed-to-empower-communities-236124 (Accessed 22 July 2025).

Tierney, K. (2018). Disaster as social problem and social construct. Chapter. In A. Javier Treviño (Ed.), *The Cambridge handbook of social problems* (pp. 79–94). Cambridge University Press, 2018.

Tiller, S., Rhindress, A., Oguntola, I., Ulku, M., Williams, K., & Sundararajan, B. (2022). Exploring the impact of climate change on arctic shipping through the lenses of quadruple bottom line and sustainable development goals. *Sustainability, 14*, Article 2193. https://doi.org/10.3390/su14042193

Verstegen, P., & Campbell, R. (2023). *The new Safeguard Mechanism and the Santos Barossa gas project*. The Australia Institute. Available at: https://australiainstitute.org.au/wp-content/uploads/2023/05/P1392-Barossa-Costs-Under-Safeguard-Mechanism-WEB.pdf

von der Porten, S., & de Loë, R. C. (2013). Collaborative approaches to governance for water and Indigenous peoples: A case study from British Columbia, Canada. *Geoforum, 50*, 149–160. https://doi.org/10.1016/j.geoforum.2013.09.001

Wang, W., & Aporta, C. (2024). Arctic marine shipping development and governance in Canada: A historical overview. *Marine Policy, 160*, Article 105958. https://doi.org/10.1016/j.marpol.2023.105958

Watson, I. (2012). Settled and unsettled spaces: Are we free to roam? In A. Moreton-Robinson (ed), *Sovereign subjects: Indigenous sovereignty matters*. Routledge.

Williamson, B. (2021). *Cultural burning in New South Wales: Challenges and opportunities for policy makers and aboriginal peoples* (Working Paper 139/2021). Centre for Aboriginal Economic Policy Research. Australian National University.

Youngblood, C. (2019). Put your money where their mouth is: Actualizing environmental justice by amplifying community voices. *Ecology Law Quarterly, 46*(2), 455–484. https://doi.org/10.15779/z383j3919j

Part IV

Resilient Farming Communities

14

Dryness of Lake Urmia in Iran and Its Impact on Farmers Displacement

Amin Sharifi Isaloo

Introduction

Lake Urmia, one of the world's largest salt lakes, the largest in Iran and the second-largest in the Middle East, has shrunk to a mere tenth of its former size (Dehghanipour et al., 2020; Kalantari et al., 2020). It is located in northwest Iran and had a rich ecosystem and beaches full of tourist centres and hotels. Today, while concerns about the effects of climate change and the increasing arid conditions of Lake Urmia have been raised since the 1970s, the costs of these changes to the surrounding environment and local communities in the region are a newer concern (SalamatNews, 2023, August 13). Salt storms caused by this dryness are an emerging threat to millions of people in north-western Iran. This adds to farmers anxieties about the future, the aggravation of drought conditions and increase in their devastating effects.

A. S. Isaloo (✉)
Department of Sociology and Criminology, University College Cork, Cork, Ireland
e-mail: amin.sharifiisaloo@ucc.ie

T. Skillington and A. Setti (eds.), *Climate Change Resilience Across Societal Contexts*, Energy, Climate and the Environment,
https://doi.org/10.1007/978-3-031-95020-9_14

According to the Environmental and Land Suitability Assessment in Iran, the farmlands of Lake Urmia basin are responsible for more than 8% of the total food produced in Iran, feeding more than seven million people (Feizizadeh et al., 2022). The drying up of Lake Urmia causes alarm not just about a lack of water and drought, but also disease, forced migration and other environment-related problems such as threatening the habitats of shrimp, flamingos, wild animals, polluting nearby cities and farms, and increased soil salinity which reduces agricultural and orchard crops yields grown around the lake. As migration from villages and small farms and/or the depopulation of many villages due to the effects of climate change, droughts, etc., is a hot debate topic in Iran, this chapter will focus on the current impacts and future risks of the drought of Lake Urmia and its links to migration or displacement of farmers and the importance of responding to these issues to effective climate resilience building. Drawing on Ulrich Beck's account of the world 'risk society', this chapter will focus on Lake Urmia in Iran and its surrounding places to demonstrate how climate change together with wrong policies can lead to displacement of farmers.

Risk Society

Beck (2005: xii) employs a new critical theory, which he also refers to as a self-critical theory to distinguish the national outlook from a cosmopolitan outlook on a globalised world and to illustrate that only a cosmopolitan outlook on climate change and its effects, with its commitment to reality, can reveal the disasters and risks that threaten our contemporary societies. 'Critical theory inquiries into the contradictions, dilemmas and unseen, unintentional side-effects of a modernity that is becoming more cosmopolitan; it draws its critical definitional power from the tension between political self-description and social scientific observation of the same' (ibid: 33). Thus, according to Beck (1992), risk society is a society increasingly preoccupied with managing and mitigating risks which are the consequences of modernity and results of environmental, technological and social changes. Beck's arguments show that modern societies are in liminal time and space shifting from an industrial stage to a risk society.

In the industrial society, the expansion of capitalism and its roots, which created issues such as poverty and inequality, led the modern society to concentrate on the production of goods and the division of labour. In contrast, the risk society pivoted on the distribution of hazards and the management of risks generated by the ongoing advancement of industrial modernity. In other words, unlike Marx and Engels, who often analyse the environmental impacts of capitalist development in the modern time in terms of division and interaction between agricultural and non-agricultural industry (Burkett, 1999), and unlike Weber and Durkheim, who discuss the relation between premodern societies and their natural environments, Beck focuses on the pervasive presence of risks and on understanding of the spread of industrial modernity and its mastery over nature rather than contradictions, economic inequality, violent confrontations, class struggles or systemic institutional failure (Jarvis, 2007). He highlights the globalisation of risk to explain that in our modern society, challenges such as climate change, nuclear accidents and environmental disasters transcend national boundaries and individual control. Both Beck (1992) and Giddens (1999) argue that these global risks are complex, uncontrollable and unpredictable and they have often long-term consequences. For example, the climate change is a global risk and its impacts are not limited to national borders. Thus, it requires international cooperation and attention.

All large lakes, such as Urmia Lake, are significant freshwater resources and they provide recreational, economic and ecosystem services to millions of inhabitants worldwide (Feizizadeh et al., 2022). The Global climate change and associated environmental changes have significantly impacted Urmia Lake during the last three decades. Lake Urmia is not the only lake has faced risk, but there are other lakes, such as Lake Chad and the Sea of Galilee, are shrinking or drying up which, in turn, lead to the globalisation of risk by creating major environmental and socioeconomic problems such as water scarcity, land degradation, and food shortages (ibid).

Cobo et al.'s (2024) recent research and their measurement of carbon dioxide and methane emission from the chronically desiccating exposed sediments of the Great Salt Lake in the US illustrate that the Great Salt Lake and other saline lakes across the world likely represent a major source of anthropogenic greenhouse gas emissions. This means greenhouse gas emissions in one part of the world can contribute to rising temperatures elsewhere. In other words, this finding of Cobo et al. (2024) and Beck's (1992, 1999) analysis of risk society demonstrates the interconnectedness of environmental risks and their transboundary effects. To address these interconnected, global and complex risks, Beck argues that all individuals and institutions must work together and there is a need for global cooperation and collaboration to prepare for and respond to emerging threats, to ensure the resilience of our risk societies and to promote sustainability and social cohesion in our current globalised world.

Beck's risk society theory, as a new critical theory framework, not only helps to demonstrate how the Lake Urmia faces serious challenges due to the increasing impacts of climate change and global warming but also poor policy choices on water diversion or the mismanagement of scarce water resources over the past few decades. These issues motivate us to think, understand and address such challenges and better manage the dangerous threats they pose, including further water crisis and mass displacements in the future.

According to UNCA (2025), climate change refers to long-term shifts in temperatures and weather patterns which can be caused by both natural disasters and human activities, but since the 1800s, human activities have been the main driver of climate change, primarily due to the burning of fossil fuels like coal, oil and gas for energy, industry, transport, buildings, agriculture and land use which are among the main sectors causing greenhouse gases. The faster climate change accelerates and the longer adaptation measures are delayed, the more challenging and costly it will become to respond to climate change. (ibid). Building climate resilience requires all stakeholders and actors, such as government, communities and businesses, to have the ability to anticipate climate risks and hazards, absorb shocks and stresses and adapt and transform development trajectories in the long run. UNCC (2025) calls this 'Adaptation' which adjusts ecological, social or economic systems in response to actual or anticipated climatic stimuli and their consequences. It encompasses changes in processes, practices and structures aimed at mitigating potential damages or capitalising on opportunities presented by climate change. In essence, countries and communities must develop adaptation strategies and take actions to address the current and future impacts of climate change (UNCC, 2025).

Decline of Lake Urmia

There are four key factors that they can affect Lake Urmia's condition and its future resilience: climate change, human activities, water diversion and salinisation. Abdollahi et al. (2024) focus on the climate change, such as global warming and increasing temperatures, which decreases the amount of available water in the lake and increases evaporation rates, to show that the unsustainable management of water resources along with climate change impacts have led Urmia Lake to severely shrink since 1995. Schmidt et al. (2021) highlight human activities such as water for irrigation and urbanisation around the Lake Urmia's basin to show the pressure on the ecosystem. Alborzi et al. (2018) find water diversion and dam construction to be among the main reasons for the lake's shrinking size and plummeting the inflow of water. Khatami and Aghajani (2020) underline salinisation to demonstrate that decreasing of the water level in lake and the concentration of salt affect seriously the ecosystem, biodiversity and the livelihoods of communities. According to the Environmental and Land Suitability Assessment in Iran, about 375,000 out of 500,000 hectares of the farmlands of Lake Urmia currently contribute to food production and based on the simulation, about 29,100 hectares will be impacted by aquifer and 20,600 hectares by soil salinisation which are anticipated to increase in the following years and as the result of this, 132,420 hectares (35.4%) of the highly productive farmlands will lose their fertility and, thus, suitability for crop development by 2050 (Feizizadeh et al., 2022). Increasing soil salinity reduces the yield of agricultural and orchard crops grown around the lake (Kalantari et al., 2020). The watershed of the lake is an important agricultural region with a population of around 6.4 million people; an estimated 76 million people live within a radius of 500 km (UNEP, 2012). Water use within the Lake Urmia basin at current rates is unsustainable without loss of the lake, and the consequent environmental damage as well as damage to the surrounding population and agriculture (UNEP, 2012).

Salt storms are also an emerging threat for millions of people in north-western Iran, because of the catastrophe of Lake Urmia (Kalantari et al., 2020). In addition, land subsidence will also increase significantly and affect the infrastructure and farmlands (Feizizadeh et al., 2022). Lake Urmia is pictured from the International Space Station by NASA as it orbited 258 miles above in July 2022 and it shows that persistent drought, dam-building and agricultural activities contributed to the lake's depletion leading to excessive red algae growth and a soaring salt content. Restoration programs, returning rainfall and falling water consumption over the last several years have begun to replenish the lake (NASA, 2022). The August 2024 picture from the International Space Station orbited 260 miles above illustrates the shrinking of the lake, causing higher salt concentrations in the water, resulting in microscopic organisms turning the water shades of red and orange over the past few decades (Cowing, 2024). According to SalamatNews (2023, August 13), only eight permanent rivers flow into the lake, and now more than 50 dams have been built on those rivers, which restrict the flow of water to the lake. It is estimated that there are 13 billion tons of salt in the bed of Lake Urmia and it has lost 95% of its water in the last three decades (ibid).

Droughts have already created significant biological, economic and social problems such as unemployment, asset erosion, reduced income, inadequate nutrition and reduced risk-taking capacity and decreasing crop yields; and its future impacts on agriculture largely depend on how farmers in the region prepare for it and how they respond to droughts (Leis et al., 2021). More intense drought conditions in Lake Urmia can lead to the migration of millions of people, particularly farmers, who live nearby and depend on its water reserves.

Climate Migration

Migration from the countryside, who are usually farmers, is one of the most common results of the climate change. Confirming this, Jahangirpour and Bakhshoodeh (2020) argue that abandoning a large number of villages in Iran during sequential droughts in recent decades is among the available evidences. If the lake dries up completely, millions of residents of the cities of West and East Azerbaijan province, including Tabriz and Urmia cities who live on the shore or in the vicinity of the lake, are under direct threat (SalamatNews, 2023). According to VOA News (2024), experts have already estimated that between six and fifteen million people affected by this incident will face a wave of respiratory diseases, forced migration and other environmental problems. If even 20% of this minimum six million population will be forced to migrate from this region, we will face about more about 1200,000 climate migrants.

Jahangirpour and Bakhshoodeh's (2020) study of the relationships between climate variables including mean temperature, annual precipitation, agriculture added value and rural migration of Iranian provinces from 2002 to 2013 indicates that rural migration increases as result of low precipitation and drought. They illustrate that the mean annual temperature and global warming, due to their negative effect on yields of agricultural products, increase migration. In other words, the effects of climate variables on added value of agriculture sector are similar to those on migration; and the added value of agricultural sector is negatively related to migration (ibid). In brief, their findings show that climate change is a main factor affecting rural migration.

According to Javad Heydarian, who is a Journalist and environmental activist, it maybe the northern half of Lake Urmia can be revived, but there is no certainty about the southern half. If this lake dries up, we should expect a crisis in the field of health and internal immigration which can affect the entire system (VOA News, 2024). Referring to the calculation of the Pistachio Growers Association, Hydarian outlines that the total value of pistachios exported in recent decades was less than the value of the water used to produce them, which means that the underground water sources were also destroyed. The lake of Urmia and other lakes in our planet are essential source of food production, nature and life. The 2022/2023 statistic shows Iran in second place of the export volume of pistachios worldwide.[1] This means that it should be considered as a problem in both local and global level and all international community should work together to prevent the dryness of the Lake Urmia.

There has been a considerable incorrect or unwise investment. This, together with waste and misuse of a huge amount of water, led to the dryness of the Lake Urmia. Therefore instead of pistachio farmers or other farmers, the policymakers and authorities should be blamed because of distorting prices which has caused unjustified investment to become justified investment and vice versa (ibid) and consequently to migration of farmers and villagers to urban areas.

Abaszadeh et al. (2020) report that the crisis in water resources and the decline of jobs related to them in villages have led to an obvious consequence: agricultural land leases. This has resulted in an increase in migration to cities, while people and private companies from urban areas who are interested in buying and renting properties and building villas within agricultural lands, farms, and gardens are increased in rural areas. This has, in turn, sharply increased private developers and small ownerships in the region and the consumption of water resources for recreational and luxury purposes (ibid).

The Perspective of Farming Communities in the Lake Urmia Region

The water crisis in any part of the world challenges the social and economic structures of agriculture and farmers. Therefore, farmers' perceptions are key to understanding issues and finding alternative solutions for managing farmers resilience. To understand the causes, contexts and consequences of the water crisis phenomenon in the Lake Urmia basin, Abaszadeh et al. (2020) examine the perception of farmers who live near the lake by conducting in-depth interviews with farmers from different social strata. The narratives of the farmers show that the core issue of the water crisis is the collapse of the order existed in their past. Other issues and concerns that their narratives indicate are the government, which is replaced with the local government, and its inability in controlling new situation, sharing existing waters or caring for farmers' water rights, bureaucracy and corruption. Farmers not only found the government responsible and accountable for any disruption and shortages, but they also believe that the government's policies and wrong decisions together with its control over natural resources have restricted their individual activities (ibid: 1).

The drought of Lake Urmia caused a water crisis in the region, leading to changes in cultivation patterns and agricultural land use. This demonstrates the importance of integrating farmers' perceptions of water and drought into policymaking (Leis et al., 2021). Based on the farmers' narratives, the shift in cultivation patterns is driven by the expansion of profitable crops and products catering to the market of industries established around the Lake Urmia basin. This has resulted in production exceeding local livelihood needs, disregarding the potential of the land and water resources. For instance, increased government incentives have encouraged the cultivation of water-intensive crops like sugar beet, which has experienced significant expansion in the past two decades. However, the government recently started discouraging planting sugar beet without considering farmer's perceptions. Explaining this, Abaszadeh et al. (2020) add direct quotations of the farmers to illustrate issues that the local farmers face and feel. I summarise a few of them here to illustrate the farmers' perceptions.

> The government came and said, Plant beets, the money is good, I'll buy them from you. We planted beets too, everyone. Planting... Now they say don't plant them, they use a lot of water. (Shirvan)
>
> Changing land use patterns and converting rain-fed agricultural lands to irrigated lands, increasing orchards, fragmenting them, and building villas are other signs of the commodification of agriculture in the (Lake Urmia) basin. (Mehdi)
>
> The share of water that was for the land is not enough anymore. You must dig a well. Every year one of them dries up and you must dig another one…Another change of use is the conversion of gardens into small villas. (Jahan)

Our village has been almost 90 percent destroyed. It is filled with villas of one hectare or even less. Who are the customers? Doctors and engineers, rich people from Tabriz and Urmia buy up the land and build villas. My father had six hectares of land and has sold four of his hectares to three people; because there is no real profit in farming. (Morad)

There is injustice in the distribution of water. Let me show you, how many people have land that is not one-third of my land; but we all have the same rights, eight hours a week. You tell me what to do when it is not enough. What should I do if I don't dig a well? (Mehdi)

The government has brought disaster to the lake. Two officials cannot agree on a law. They don't know what they are doing...They don't tell us what they are doing. I don't trust any of them; they are not honest. (Afshin)

During elections, this representative comes, that councillor comes, We will give you your rights, we will increase your water quota by this much... it's all a lie. (Merdad)

You and I, this village, and ten other villages cannot revive the lake. Everyone must join hands... No one is willing to use their share of water to go to the lake (Ahmad).

Our Children have the right to see and enjoy this nature. They have the right to continue their farming business if they want. Perhaps saving money for agriculture would be a good idea; because a plant is meaningless without water' (Hekmat).

Agriculture means water. They say that agriculture has brought drought. What can a poor farmer do? He waits all year for the harvest, his wheat is in great pain, he is afraid of it, why should he buy it at such a high price?

My observation of the secondary data and content analyses shows that the farmers and the Iranian authorities blame each other for the water crisis and dryness of the Lake Urmia basin during the last decade. Some of the government officials even suggested to inform the farmers and let them know about their faults and their usage of water, wells and underground water. They have not clearly discussed the sanctions of the West imposed on Iran last few decades, and the global risks and issues of the water crises and dryness of the lake.

The following might help to address poor policymaking on water management to improve long-term resilience:

1. Awareness raising, and cultural and educational Developments (Hadipour et al., 2024),
2. Addressing water supply imbalances,
3. Controlling dust,
4. Measuring evapotranspiration regularly,
5. Avoiding degradation of ecosystem,
6. Monitoring underground water,
7. Mobilising resources,
8. Monitoring agricultural leasing and water usages,
9. Considering conservations and costs,
10. Carry out climate risk assessments at national, local (UNCC, 2025),
11. Creating data hub development which can monitor and track progress, share knowledge, experiences and solutions to provide consistent access to data (ibid),
12. Building climate resilience.

The Resilience Strategies of Farmers

In response to the drought and the shortage of water in Lake Urmia, the Restoration Headquarters was established in 2013 in order to study and provide funding to reduce water wastage and try to return at least some water to the lake which in turn could help farmers resilience and climate migration. In the same year, they approved more than eighty plans in the basin, among which are the release of water from dams and the creation of a channel for water to enter Lake Urmia (News, 2023; Salamat Khatami & Aghajani, 2020). A long-delayed project, designed to divert water from the Kani Sib dam in West Azerbaijan province, finally began transferring water in 2023 (SalamatNews, 2023). Local communities and governments have also introduced adapting to the changing environmental conditions such as exploring alternative livelihoods and water conservation practices to encounter droughts (Rashed & Sadeghian, 2021). To reduce the amount of water being diverted from the rivers, Iranian authorities began to reduce water usage in agriculture and improve irrigation efficiency (Rashed & Sadeghian, 2021). The other effort for resilience and restoration of the lake have been international support and research, particularly on sustainable water management and restoring the natural ecosystem (UNEP, 2017).

The efforts of the construction of smaller dams and the release of water into the lake for improving the ecological balance of the Lake Urmia and the rains at the end of July and the beginning of August 2023 caused relatively optimistic assumptions to provide part of the country's lack of rainfall and provide part of the water of Iran's dried up lakes, but a significant amount of water did not contribute to the catchment area of Lake Urmia and practically did not change the condition of the lake and its surroundings lands (News, 2023; Salamat Khatami & Aghajani, 2020). This increased farmers' anxieties about the future of drought and its devastating effects. Some strategies such as the use of advanced irrigation systems, drought-adapted crop cultivation, storage and optimisation of consumption, getting help from religious institutions and migrating to the city were considered to combat issues caused by drought or to reduce the effects of drought (Leis et al., 2021). Another popular strategy for bringing additional water was cloud seeding attempting to increase precipitation by dispersing substances into clouds, but cloud seeding in general is controversial and its impact limited making this a partial and uncertain solution (UNEP, 2012).

Iran has also provided large government loans for the agricultural sector to expand and switch from being primarily rain-fed to irrigated under its five-year economic development programme. To bring the necessary water for the farms, as well as for growing domestic and industrial use, Iran has constructed more than 50 dams on rivers that drain much of north-western Iran and flow into the Lake Urmia (Kalantari et al., 2020). While these dams siphoned off the water that once fed the lake, the drying process was intensified by climate change. The rate of rainfall has reduced in recent decades and the Urmia basin has experienced several multi-year droughts (ibid).

To enhance the resilience of Lake Urmia, Iran's government has initiated several water management schemes such as prohibiting further water withdrawal for irrigation, which encountered farmers' reactions as indicated in last page, temporary suspension of dam construction, change of cropping patterns and developing alternative water resources (ULRP, 2015). But transformation of these programs into action has been slow, and their adoption rates have been low among the farmers of the Lake Urmia Basin. Furthermore, several carbon-smart, energy-smart, knowledge-smart, nutrient-smart and water-smart interventions have been developed for making agricultural systems climate-smart, but the efficiency of these technologies and practices relies on the socio-ecological context of the target regions. Nevertheless, the potential impacts of climate smart interventions on restoration of Lake Urmia and its surroundings lands have not been well documented (Maleki et al., 2022) and as scenario-based results of the study of Feizizadeh et al. (2022) indicate, the Lake Urmia basin is going to face critical environmental conditions.

The problem with all these strategies is that they are designed and considered based on the national outlook rather than the cosmopolitan one. In our globalised planet and so-called 'modern time', 'the cosmopolitan outlook opens up spaces and strategies for action that the national outlook closes off' (Beck, 2005: 33). Thus, for the this type of global risks and problems, we need to take a global perspective and act globally.

Discussion

Farmers in the Lake Urmia basin region believe that water resources have infiltrated social capital and deeply ingrained moralities, leading to distrust in interpersonal and social relations. This mistrust extends to political institutions, government support, and each other, causing uncertainty about their future (Abaszadeh, 2020: 2). Although the role of human activities and government policies in the dramatic decline of the Lake Urmia's water should not be ignored, the modernity, climate change and environment issues must be discussed seriously at the international level, as much as the national level. These are risks that Beck (1992) has critically discussed by focusing on the pervasive presence of risks in late industrial modernity and by highlighting the globalisation of risk to explain that in our modern society, challenges such as climate change and environmental disasters are complex global risks as much as they are national, local and individual concerns. The dryness of the Lake Urmia has long-term consequences in ways with global and local relevance.

On the one hand, the 2017 reports of the United Nations Environment Programme (UNEP) provide a global perspective on Lake Urmia's environmental issues, addressing international efforts and recommendations for long-term solutions. On the other hand, AghaKouchak et al. (2015) note that the impending collapse of Lake Urmia's unique ecosystem requires immediate international attention in order to prevent the catastrophic socio-ecological repercussions. But Madani's (2020) study obviously shows that sanctions, particularly economic sanctions, impact the environment and lower the priority of the environmental sector in the public policy agenda, and increase environmental issues in practice. Iran has been subject to several sanctions since the 1979 revolution. The diversity and multiplicity of these sanctions have seriously affected Iran's economy, as well as capacity to allocate sufficient resources to addressing mounting environmental challenges. It limited access to international markets to buy necessary tools and equipment and to collaborate with international institutions and scientists to tackle the dryness of the Lake Urmia which, arguably, is a collective risk to human and non-human communities in need of international regulation and more dedicated water resilience-building policies. The challenge is to devise resilience-building strategies that work with the current reality of unreliable and highly variable water sources and develop new efficient water conservation measures (reducing the amount of water needed to grow crops, irrigate farms, private and commercial water consumption, etc. See Rockström, 2003: 869).

Both modern risks and 'developments are too much for individual states to deal with on their own' because of not having appropriate ways on intervening and observation (Beck, 2005: 32). Thus, extensive research and global collaboration are necessary to develop a comprehensive restoration plan. To this end, involvement of international organisations can foster a creative dialogue on how to address this environmental crisis and call for action to develop sustainable restoration ideas and to put new visions and strategies into practice (Madani, 2020).

Conclusion

Based on Beck's arguments, it can be concluded that the water crisis phenomenon in the Lake Urmia basin is the consequences of environmental, technological and social changes. The total loss of the lake would be tragic for Iran and other parts of the world in many ways, from the flamingos that feed on Artemia (unique crustaceans that live in the lake's salty waters), to the loss of biodiversity and damage to agriculture in the Lake Urmia catchment (SalamatNews, 2023). In addition, the minerals and sediments on the bottom of the lake contain not only salt, but also heavy toxic metals that are used in industry and agricultural poisons. As the lake bed is exposed and dried up, these materials have become dust in the air, posing a serious threat to the environment and people (ibid).

Considering that 'the space for action opened up by the cosmopolitan outlook stands in the contradiction to the lack of alternatives diagnosed by politicians and social scientists alike in the national perspective' (Beck, 2005: 33), the international institutions and working groups should not only monitor this global risk, but also the performance of governments in the lake basin and its surroundings such as lands and farms. The problem of the Lake Urmia is a global risk and should be discussed in global level to deal with its current situation, to prevent further dryness and to increase farmers resilience to drought which, in turn, can help to prevent and/or reduce the farmers displacement and migration to cities and abroad. Any global risk is a problem which affects the world as a whole, thus, all global risks should be exempted from any type of sanctions that superpowers adopt to enforce their own preplanned interests and international politics. The global risks such as the climate change, drought and related environment issues must be tackled urgently on a global level as they concern the foundation of our existence.

Note

1. https://www.statista.com/statistics/933015/pistachio-exports-global-by-country/.

References

Abaszadeh, A., Banifatemeh, H., & Nikdel, N. (2020). A study of farmers' perception of Lake Urmia (UL) crisis with grounded theory approach. *Journal of Applied Sociology, 31*(2), No. 78, 1–22.

AghaKouchak, A., Norouzi, H., Madani, K., Mirchi, A., Azarderakhsh, M., Nazemi, A., Nasrollahi, N., Farahmand, A., Mehran, A., & Hasanzadeh, E. (2015). Aral Sea syndrome desiccates Lake Urmia: Call for action. *Journal of Great Lakes Research, 41*(1), 307–311.

Alborzi, A., Mirchi, A., Moftakhari, H., Mallakpour, I., Alian, S., Nazemi, A., Hassanzadeh, E., Mazdiyasni, O., Ashraf, S., Madani, K., Norouzi, H., Azarderakhsh, M., Mehran, A., Sadegh, M., Castelletti, A., & AghaKouchak, A. (2018). Climate-informed environmental inflows to revive a drying lake facing meteorological and anthropogenic droughts. *Environmental Research Letters, 13*(8), 1–12.

Beck, U. (1992). *Risk society: Towards a new modernity*. Sage Publications.

Beck, U. (1999). *World risk society*. Polity Press.

Beck, U. (2005). *Power in the global age: A new global political economy*. Polity Press.

Burkett, P. (1999). *Marx and nature: A red and green perspective*. St. Martin's Press.

Cobo, M., Goldhammer, T., & Brothers, S. (2024). A desiccating saline lake bed is a significant source of anthropogenic greenhouse gas emissions. *One Earth, 7*(8), 1414–1423.

Cowing, K. (2024, August 23). *Orbital biosignature survey of Lake Urmia*. NASA. https://astrobiology.com/2024/08/orbital-biosignature-survey-of-lake-urmia.html. Accessed December 10, 2024.

Dehghanipour A, H., Panahi, D, M., Mousavi, H., Kalantari, Z., & Tajrishy, M. (2020). Effects of water level decline in Lake Urmia, Iran, on local climate conditions. *Water (MDPI), 12*, 2153, 1–15. https://doi.org/10.3390/w12082153.

Feizizadeh, B., Lakes, T., Omarzadeh, D., Sharifi, A., Blaschke, T., & Karimzadeh, S. (2022). Scenario-based analysis of the impacts of lake drying on food production in the Lake Urmia Basin of Northern Iran. *Scientific Reports, 12*(6237), 1–16. https://doi.org/10.1038/s41598-022-10159-2

Abdollahi, Z., Feizizadeh, B., Shokati, B., Gaiolini, M., Busico, G., Mastrocicco, M., & Colombani, N. (2024). Simulating the effects of retreating Urmia Lake and increased evapotranspiration rates on the nearby unconfined aquifer. *Groundwater for Sustainable Development, 26*, 1–11.

Giddens, A. (1999). *Runaway world: How globalization is reshaping our lives*. Profile Books.

Hadipour, M., Pourebrahim, S., Heidari, H., Nikooy, F., Ahmed, A. N., & Ern, C. J. (2024). Evaluation of water resource balance in the Urmia Lake Basin: Integrating carrying capacity and water footprint model for sustainable management. *Ecological Indicators, 166*, 1–14.

Jahangirpour, D., & Bakhshoodeh, M. (2020). Panel VAR (the impact of climate change on rural migration in Iran: Application of panel VAR approach). *Environmental Researches, 11*(21), 133–142.

Jarvis, D. S. L. (2007). Risk, globalisation and the state: A critical appraisal of ulrich beck and the world risk society thesis. *Global Society, 21*(1), 23–46.

Kalantari, Z., Panahi, D, M., & Destouni, G. (2020, August 3). Iran: decades of unsustainable water use has dried up lakes and caused environmental destruction. *The Conversation*. https://theconversation.com/iran-decades-of-unsustainable-water-use-has-dried-up-lakes-and-caused-environmental-destruction-140565. Accessed December 10, 2024.

Khatami, R., & Aghajani, M. (2020). Restoration and management of Lake Urmia: Challenges and strategies. *Iranian Journal of Environmental Health Science & Engineering, 17*(1), 51–58.

Leis, N., Rostami, F., & Beige A., H., A. (2021). Analysis of farmers' perception of drought in Lake Urmia Basin and its impacts (Case Study: Miandoab County). *Human Geography Researches, 53*(3), 957–976. https://jhgr.ut.ac.ir/article_76758_c6db88dc3298b8174e25e8379521bf35.pdf?lang=en, Accessed November 25, 2024.

Madani, K. (2020). How international economic sanctions harm the environment. *AGU, 8*(12), 1–12.

Maleki, T., Koohestani, H., & Keshavarz, M. (2022). Can climate-smart agriculture mitigate the Lake Urmia tragedy in its eastern basin? *Agricultural Water Management, 260*, Article 107256. https://doi.org/10.1016/j.agwat.2021.107256

NASA (2022, August 15). *Lake Urmia in northwest Iran*. https://www.nasa.gov/image-article/lake-urmia-northwest-iran/. Accessed November 22, 2024.

Rashed, E. A., & Sadeghian, S. (2021). Socio-environmental resilience in the face of Lake Urmia's decline: A study of adaptation strategies in local communities. *Environmental Science and Policy, 118*, 34–43.

Rockström, J. (2003). Resilience building and water demand management for drought mitigation. *Physics and Chemistry of the Earth, Parts a/b/c, 28*, 869–877.

SalamatNews (2023, August 13). Wave of migration following the drying up of Lake Urmia. *Health news agency*. https://www.salamatnews.com/news/359221/موج-مهاجرت-ها-در-پی-خشک-شدن-دریاچه-ارومیه. Accessed November 25 2024.

Schmidt, M., Gonda, R., & Transiskus, S. (2021). Environmental degradation at Lake Urmia (Iran): Exploring the causes and their impacts on rural livelihoods. *GeoJournal, 86*, 2149–2163.

ULRP (Lake Urmia Restoration Program 2015). *Lake Urmia Restoration plan: Challenges and solutions*.

UNCA (2025). *United Nations climate action: What is climate change?* https://www.un.org/en/climatechange/what-is-climate-change Accessed March 10, 2025.

UNCC (2025). United Nations climate change: Adaptation and resilience. https://unfccc.int/topics/adaptation-and-resilience/the-big-picture/introduction, Accessed March 10, 2025.

UNEP (United Nations Environment Programme) (2012, February). *The drying of Iran's Lake Urmia and its environmental consequences*. https://na.unep.net/geas/getUNEPPageWithArticleIDScript.php?article_id=79. Accessed December 2, 2024.

UNEP (United Nations Environment Programme) (2017). Lake Urmia: Signs of recovery. *Foresight Brief*. https://wedocs.unep.org/bitstream/handle/20.500.11822/22312/Foresight_%20Brief_%20004_2017.pdf?sequence=1&isAllowed=y

VOA News (2024, November 25). https://ir.voanews.com/a/iran-urmia-lake-crisis-getting-worse/7226441.html; https://ir.voanews.com/a/iran-water-crisis-drought/7218564.html. Accessed December 5, 2024.

15

Aran Islands: The Challenge of Maintaining Grassland Biodiversity in Future Climate Change-Led Policy Structure

Patrick McGurn

Introduction

The Aran Islands, Inis Mór, Inis Meáin and Inis Oírr, are located at the mouth of Galway Bay, on the West coast of Ireland. Agriculture has shaped their landscape over the last six thousand years. Neolithic farmers arrived, removing the original tree cover to expose the limestone rock and grasslands. Future generations of farmers carried sand and seaweed to make the soil on which many of the species-rich grasslands are found. Thousands of kilometres of stone wall were built to shelter their livestock from the harsh Atlantic winds, to protect the grasslands, mark boundaries and as a way of storing the stone gathered from the fields. Out of necessity an agricultural system developed. The history of this landscape is still very evident today.

P. McGurn (✉)
ACRES Burren Aran, Ennis Co Clare, Carron, Ireland
e-mail: patrick@acresburrenaran.ie

T. Skillington and A. Setti (eds.), *Climate Change Resilience Across Societal Contexts*, Energy, Climate and the Environment,
https://doi.org/10.1007/978-3-031-95020-9_15

Aran Islands—the Farming System

The Aran Islands are approximately 43.3 km2 or 4330 ha in size. Their geology is mainly karst limestone dating from the Viséan age of the Lower Carboniferous, formed as sediments in a tropical sea approximately 350 million years ago. Human involvement with the Aran Islands dates to at least the Middle Stone Age (Mesolithic—between 6 and 10 thousand years ago) when they were visited by hunter-gatherer groups coming from the coast of Clare. Continuous inhabitancy and natural forces have all shaped the appearance of the islands to its present form. Tim Robinson (1986) in his book Stones of Aran describes the islands as an 'incredible jigsaw puzzle of little fields where farmers clear their stony patches and mark their every increasing subdivision of their holdings by building walls'. Mc Gurn et al., (2020) describe the agricultural system which has developed as encompassing cattle grazing part of the farm during the growing season, while the remainder is left to allow excess grass to grow that can then be grazed in the winter as a standing crop. Beef cows are the main enterprise with calves sold to the mainland for finishing. The summer grazing tends to be on deeper soils, grazed from calving time (March/April) to late Autumn/Early winter. The winter grazing is left ungrazed during this period to build up a bank of grass for grazing from November to March/April. Excess grass during the summer grazing was traditionally turned into hay, allowing some fodder supplementation in the winter, but this practice has declined and hay is now purchased from the mainland. This winter age practice was likely to have been widespread in Ireland at one stage and in agricultural terms, the standing crop is known as foggage. However, this system has changed over the years due to the ascendancy of hay and then, silage conservation. This farming system today is mainly limited to the Aran Islands and the Burren, where the drier limestone grasslands and limestone pavement are less prone to poaching. It is an efficient farming system; no cattle housing or slurry storage is required and the system takes full advantage of compensatory growth (animal growth may be lower than expected for some months due to under-nutrition; later, the liveweight gain of the cattle will be greater than expected due to good nutrition in the available forage) (Mc Gurn et al., 2020). Characteristic of the system is a low stocking rate and high labour requirements with low financial returns. The system has been in practice over many years, missing the intensification of agriculture seen elsewhere. In 1684, Ruaidhrí Ó Flaithbheartaigh, the Irish historian and 'de jure Lord of Iar-Connacht', wrote in his book 'A Chorographical Description of West Or H-Iar Connaught': 'The soil is almost paved over with stones, so as in some places nothing is to be seen but large stones with wide openings between them where cattle break their legs. Scarce any other stones there but lime stones and marble fit for tombstones, chimney mantle trees and high crosses. Among these stones is very sweet pasture so that beef and mutton are better and earlier in season here than elsewhere; and of late there is plenty of cheese & tillage' (O'Flaherty, 1684).

Island Ecology

As well as food production, the islands' agricultural system has resulted in a high species diversity of flora, intact historic landscapes and cultural heritage throughout the islands (O'Rourke, 2006).The islands contain 17 different habitats types that are increasingly rare in Europe and listed in the EU Habitats Directive. These include Coastal lagoons (1150*), Fixed coastal dunes with herbaceous vegetation (grey dunes) (2130*), Machair (21AO*), Semi-natural dry grasslands and scrubland facies on calcareous substrates (Festuco-Brometalia) (important orchid sites) (6210*) and Limestone pavement (8240*), Reefs (1170), Perennial vegetation of stony banks (1220), Vegetated sea cliffs of the Atlantic and Baltic coasts (1230), Embryonic shifting dunes (2110), Shifting dunes along the shoreline with Ammophila arenaria (white dunes) (2120), Dunes with Salix repens ssp. argentea (Salix arenariae) (2170), Humid dune slacks (2190), European dry heaths (4030), Alpine and Boreal heaths (4060), Lowland hay meadows (Alopecurus pratensis, Sanguisorba officinalis) (6510), Turloughs (3180*), Submerged or partly submerged sea caves (8330). Six of these habitats are classed as priority habitats (*). Based on Halada et al. (2011), nine of the 17 habitat types are fully or partly dependent on agricultural management. This is reflected in the National Parks and Wildlife Service (NPWS) site synopsis that the islands are of 'major scientific importance owing to the range of outstanding karstic carboniferous limestone and coastal habitats, and the number of rare and threatened species found thereon. The cultural heritage of the islands (and in particular the continuation of traditional low-intensity farming practices) is intrinsically linked with its scientific interest' (NPWS, 1997). The main habitat types found are the Semi-natural dry grasslands and scrubland facies on calcareous substrates (Festuco Brometalia) (important orchid sites) (6210*) and Limestone pavement (8240*). These two habitat types form a mosaic across the islands and are the main focus of the farming system. While the biodiversity of the Aran Islands is high, the annual dry matter yields are low, ranging from 926 kg DM ha-1 yr-1 to 6704.1 kg DM ha-1 yr-1 in the more fertile sites (Duignan et al., 2018). Ensuring low fertility is an important aspect in maintaining species diversity (Plantureux et al., 2005).

HNV Farming

The Aran Islands are part of Ireland's High Nature Value Farmland (HNV). HNV farmland is associated with areas where agriculture is the major land use and where low intensity agricultural systems support or are associated with high levels of semi-natural vegetation that support species and/or habitats of conservation concern. (Beaufoy et al., 1994, 2008). In Ireland, HNV farmland occurs across a broad range of landscape type, covering one third of the agricultural area (Matin et al., 2020) equating to approximately 1.5 million ha of agricultural land. These areas may have natural constraints on food production related to soils, topography, climate and remoteness but are associated with high levels of biodiversity, landscape and socio-cultural values (Paracchini & Oppermann, 2012), or in other cases have bypassed the intensification process (Lomba, 2013). In recognition of their high nature value, approximately 50% of these areas are part of the Natura 2000 network (Matin et al., 2020).

HNV farmland has an important role in maintaining Ireland's Biodiversity. For example, the Aran Islands contain over 60% of Irelands plant species found in Ireland (Webb, 1980) and so are one of Irelands biodiversity hotspots and a genetic seedbank of Irish flora. Recent agri-environment projects have recognised the importance of the islands for their biodiversity services. The AranLIFE project, an EU co-funded LIFE project and the Caomhnú Árann, a European Innovation Partnership, have worked with farmers addressing the issues reducing the biodiversity value of the islands grasslands and rewarding farmers for the production of higher quality species-rich grasslands using a grassland scoring system. Results from these projects have been incorporated into Irelands mainstream agri-environment scheme, Agri-Climate Rural Environment Scheme (ACRES), specifically targeting the Burren/Aran area with a local project team.[1]

In Ireland, figures outlining our biodiversity loss are stark with many plant, animal and bird species in danger of extinction, almost 30% of semi-natural grasslands have disappeared, more than half of the countries marine environment is in an unhealthy state and almost 50% of our freshwater systems are in unsatisfactory condition. (DCHG, 2019). According to The Plant Atlas 2020, more than half Irish native plant species are in decline, with native grasslands suffering the largest decline (BSBI, 2020). Due to declines in Ireland's Biodiversity in May 2019, the Dail voted upon and declared both a National Climate and Biodiversity Emergency. Following from this, the Irish Government has developed a National Biodiversity Action Plan (Government of Ireland, 2023) and a Climate Action Plan 2024.

In relation to the Climate Action Plan, the Climate Action and Low Carbon Development Act commits Ireland to a legally binding targets of a climate neutral economy no later than 2050, and to a reduction in emissions of 51% by 2030 (compared to 2018 levels). The key driver of agricultural GHG emissions in Ireland continues to be the level of activity in bovine agriculture (Lanigan et al., 2023). The research behind proposed reduction measures is that the Irish agriculture sector can achieve the set mitigation target by adapting livestock production systems, resulting in agricultural efficiency gains, and by implementing specific mitigation technologies (Adenaeuer et al., 2022). This intensive approach is geared towards maximising production to deliver higher outputs with associated lower emissions. To achieve the emission reduction targets, the agriculture sector requires a combination of high levels of measure adoption, limiting animal number increases and the development of new science. Three future scenarios for livestock systems are suggested by Teagasc, Scenario 1 is the Base case (mid), S2 the low scenario and S3 the high scenario. All suggest a reduction of 29%, 49% and 26% respectively in beef cow numbers, with only S1 and S2 achieving the 2030 emissions reduction target and under the carbon budget target (Lanigan et al., 2023). However, to be truly viable, a target-based system ought to consider livestock numbers in context and in particular, in relation to geographical, social, cultural and economic factors that can vary considerably across regions. At present, policy does not show sufficient sensitivity to these variables and for such reasons, is unlikely to contribute to a green transition that is genuinely sustainable and equitable from the perspective of all those concerned.

Climate Change Policy and the Islands.

To maintain the diversity of the Aran Island grasslands, grazing with livestock is required. Calcareous grasslands tend to be low productivity systems, which produce low yields of digestible herbage, and so they are usually maintained by grazing rather than mowing. Grazing plays a key role in maintaining species richness by limiting the ability of competitive species to achieve dominance. It is also the preferred option when managing for invertebrates (Olmeda et al., 2019). Dunford (2002) found that without grazing animals on Burren calcareous grasslands, there is a loss in plant species diversity that results from the under grazing and neglect of these grasslands and the discontinuation of agricultural activity is perhaps the greatest future threat to the species-rich upland grasslands of the Burren. Smith et al. (2010) found traditional grazing regimes are likely to be the most effective and sustainable way to maintain, and in some instances restore, the biodiversity of semi-natural grasslands within priority farmland habitat on the Aran Islands.

Future policy, with a reduction in beef cow numbers and agricultural greenhouse gas mitigation measures such as Reduced Beef Finishing Age,

Provision of Clover and Multi Species Swards, Mineral Soil Drainage, Liming and Increased Soil Phosphorus (Lanigan et al., 2023), is designed with a GHG reduction with food production viewpoint and many measures will have a negative biodiversity affect, thus continuing the spiral of biodiversity decline. European habitats and landscapes considered important for biodiversity conservation depend on the continuation of specific farming systems (Lomba et al., 2023). Across the upland rangelands of northern and western Europe, extensive suckler cow production is important in utilising semi-natural pastures (Morgan-Davies et al., 2014). Such extensive systems play a dual role of both food production and of upholding rural and social landscapes and maintaining these landscapes and the diversity contained within them requires the continuation of appropriate pastoral activities (Morgan-Davies et al., 2014).

As an agriculture system, the Aran Island farming system has a high level of resilience. Meuwissen et al. (2019) define resilience of a farming system as its ability to ensure the provision of the system functions in the face of increasingly complex and accumulating economic, social, environmental and institutional shocks and stresses, through capacities of robustness, adaptability and transformability. The islands low input system, with low dependence on external inputs, has been in existence as described by Ó Flaithbheartaigh in 1684 (O'Flaherty, 1684), surviving storms, droughts, institutional shocks and stresses.

Resilience is becoming a key concept in the discussions of sustainable development and long-term viability of agricultural systems (Žickiene et al., 2022) but it is targeted more to food production than other ecosystem services provided by agriculture. Volkov et al. (2021) in assessing the Economic Resilience of Agriculture selected production of food, assurance of farm viability and provision of income as the core functions of an agricultural sector. Žickiene et al. (2022) single out the three main socio-economic functions in relation to the economic resilience of the agricultural sector: (a).the production of food and other bio-based resources at affordable prices; (b).the assurance of farm viability; (c).The creation and maintenance of decent jobs.

The European Commission has emphasised that a more resilient farming sector is required to better respond to current and future economic, societal and environmental challenges. Consequently, supporting resilience has become an important aim of the proposals of the Common Agricultural Policy (CAP) post-2020 (EC, 2018). Defining resilience purely in terms of economics, coupled with climate mitigation measures which focus on food production only, will have consequences for not only the Aran Islands but also other HNV areas throughout Europe.

The Aran Islands are a suitable case study showing that to maintain biodiversity, grazing livestock are required. The lower feeding value of the Aran Grasslands compared to reseeded pastures or Clover and Multi-species swards and associated low grassland production yields (Duignan et al., 2018) are not in line with achieving set mitigations target suggested by Adenaeuer et al., 2022. HNV farmland habitats and landscapes that are considered important for biodiversity conservation depend on the continuation of specific farming systems (Lomba et al., 2023). The specific characteristics of the Aran Islands and other HNV farmland in Ireland need to be considered when developing carbon efficiency support measures. Concentrating policies purely based on food production as an output could lead to further environmental losses. Ireland has declared both a National Climate and Biodiversity Emergency, the effect of addressing one issue, such as climate, in isolation (i.e., reducing numbers of beef cow herds) needs to be considered in terms of its contribution to further exacerbating Ireland's biodiversity crisis, as it does in other world regions as well (see, for example, Hendricks et al., 2007, Metera et al., 2011).

Notes

1. https://www.acresburrenaran.ie/

References

Adenaeuer, L., Breen, J., Witzke, P., Kesting, M., Hayden, A., & Donnellan, T. (2022). The potential impacts of an EU-wide agricultural mitigation target on the Irish agriculture sector. *Climate Policy, 23*(4), 495–508. https://doi.org/10.1080/14693062.2022.2105791

BSBI (2020). *Ireland's changing flora a summary of the results of plant Atlas 2020*.

Beaufoy, G., Baldock, D., & Clark, J. (1994). *The nature of farming: low-intensity farming systems in nine European countries*. Institute for European Environmental Policy.

Beaufoy, G., G. Jones, K. de Rijck, & Kazakova, Y. (2008). *High nature value farmlands: recognising the importance of south east European landscapes*. Final summary report (Bulgaria and Romania). WWF Danube-Carpathian Programme and European Forum on Nature Conservation and Pastoralism.

DCHG (2019). *Ireland's 6th national report to the convention on biological diversity*. Department of Culture, Heritage and the Gaeltacht

Duignan L., Ó hUallacháin D., Finn J., & Moran J. (2018). *Agricultural practices and the quality of calcareous grassland habitat on the Aran Islands*. AranLife report.

Dunford, B. (2002, December). *Farming and the Burren*. Teagasc.

EC: European Commission (2018). *Future of the common agricultural policy*. Retrieved from, https://ec.europa.eu/info/food-farming-fisheries/key-policies/common-agricultural-policy/future-cap_en#objectives

Halada, L., Evans, D., Romão, C., & Petersen, J.-E. (2011). Which habitats of European importance depend on agricultural practices? *Biodiversity and Conservation, 20*, 2365–2378.

Hendricks, H. H., Bond, W. J., Midgley, J. J., & Novellie, P. A. (2007). Biodiversity conservation and pastoralism—reducing herd size in a communal livestock production system in Richtersveld National Park. *Journal of Arid Environments, 70*, 718–727.

Lanigan G., Black, K., Donnellan T.,Crosson P.,Beausang C., Hanrahan K., Buckley C., Lahart B., Herron J., Redmond J., Shalloo L., Krol D., Forrestal P., Farrelly N., O'Brien D., Lenehan J.J., Hennessy M., O'Donovan M., Wall D., O'Sullivan L., O'Dwyer T., Dineen M., Waters S., NiFlahartha N., Houlihan T., Murphy P., Spink J., Dillon P., Upton J., Richards K. (2023). *MACC 2023: An updated analysis of the greenhouse gas abatement potential of the Irish agriculture and land-use sectors between 2021 and 2030* (p. 353pp). Teagasc.

Lomba, A., McCracken, D., & Herzon, I. (2023). Editorial: High nature value farming systems in Europe. *Ecology and Society, 28*(2), 20. https://doi.org/10.5751/ES-14159-280220

Matin, S., Sullivan C. A., Finn, J., Ó hUallacháin, D., Green, S., Meredith D., Moran, J. (January, 2020). Assessing the distribution and extent of high nature value farmland in the Republic of Ireland. *Ecological Indicators, 108.*

Mc Gurn et al. (2020). In E. O'Rourke & J. Finn (Eds.), *Farming for nature-the role of results-based payments*. Teagasc and National Parks and Wildlife Services.

Metera, E., Sakowski, T., Słoniewski, K., & Romanowicz, B. (2011). Grazing as a tool to maintain biodiversity of grassland - a review. *Animal Science Papers and Reports, 28*(4), 315–334.

Meuwissen, M. P. M., Feindt, P. H., Spiegel, A., Termeer, C. A. J. M., Mathijs, E., de Mey, Y., Finger, R., Balmann, A., Wauters, E., Urquhart, J., Vigani, M., Zawalińska, K., Herrera, H., Nicholas-Davies, P., Hansson, H., Paas, W., Slijper, T., Coopmans, I., Vroege, W., Ciechomska, A., Accatino, F., Kopainsky, B., Poortvliet, P. M., Candel, J. J. L., Maye, D., Severini, S., Senni, S., Soriano, B., Lagerkvist, C.-J., Peneva, M., Gavrilescu, C., & Reidsma, P. (2019). A framework to assess the resilience of farming systems. *Agricultural Systems, 176*(N/A), 102656. https://doi.org/10.1016/j.agsy.2019.102656

Morgan-Davies, J., Morgan-Davies, C., Pollock, M. L., Holland, J. P., & Waterhouse, A. (2014). Characterisation of extensive beef cattle systems: disparities between opinions, practice and policy. *Land Use Policy, 38*, 707–718. Advance online publication. https://doi.org/10.1016/j.landusepol.2014.01.016

National Parks and Wildlife Service (1997). *SAC site synopses*. Accessed on website: www.npws.ie/en/ConservationSites/SpecialAreasofConservationSACs/Galway/

O'Flaherty, R. (1684). *A chorographical description of West or Iar Connaught*. Irish Archaeological Society. Dublin.

Olmeda, C., Šefferová, V., Underwood, E., Millan, L., Gil, T. & Naumann, S. (compilers). (2019). *EU Action plan to maintain and restore to favourable conservation status the habitat type 6210 Semi-natural dry grasslands and scrubland facies on calcareous substrates (FestucoBrometalia) (*important orchid sites)*. European Commission Technical Report 2019.

O'Rourke, C. (2006). *Nature Guide to the Aran Islands*. Lilliput Press.

Paracchini, M. L. & Oppermann, R. (2012). Public goods and ecosystem services delivered by HNV farmland. In R. Oppermann, G. Beaufoy, &

G. Jones (Eds.), *High nature value farming in Europe: 35 European countries—Experiences and perspectives* (pp. 446–450). verlag regionalkultur, Ubstadt-Weiher.

Plantureux, S., Peeters, A., & McCracken, D. (2005). Biodiversity in intensive grasslands: Effects of management, improvements and challenges. *Agronomy Research, 3*(2), 153–164.

Robinson, T. (1986). *Stones of Aran: Pilgrimage*. Lilliput Press.

Smith, G.F., Bligh, J., Delaney, E., Egan, M., O'Donavan, G., O'Donaghue, P., & O'Hara, K. (2010). *Case studies on high nature value farming in Ireland: Aran Islands and Connemara*. A report to the Heritage Council, Kilkenny Ireland.

Volkov, A., Žiˇckiene, A., Morkunas, M., Baležentis, T., Ribašauskien˙ e, E., & Streimikiene, D. (2021). A multi-criteria approach for assessing the economic resilience of agriculture: The case of Lithuania. *Sustainability, 13*, 2370.

Webb, D.A. (1980) The Flora of the Aran Islands. *Journal of Life Sciences Royal Dublin Society, 2*, 51–83; Number of Pages: 33.

Žickiene, A., Melnikien, R., Morkunas, M., & Volkov, A. (2022). CAP direct¯payments and economic resilience of agriculture: Impact assessment. *Sustainability, 14*, 10546.

16

Advisory Support for Farmer Wellbeing and Climate Change Mitigation

John McNamara, Tom O'Dwyer, and Pádraig Flattery

Introduction

This chapter firstly outlines (1) the current structure of Irish agriculture and the role played by the Irish Agricultural Advisory Service in its development, secondly (2) it describes Ireland's climate change and biodiversity legislation, then (3) it outlines information on farmer wellbeing and its association with climate change, with particular reference to Irish studies. Fourthly, (4) it describes the overall role of the Irish Agricultural Advisory Service and focuses on its role assisting farmers

J. McNamara · P. Flattery
Teagasc—Irish Agriculture and Food Development Authority, Carlow, Ireland
e-mail: john.g.mcnamara@teagasc.ie

P. Flattery
e-mail: Padraig.Flattery@met.ie

T. O'Dwyer (✉)
Met Eireann—Irish Meteorological Service, Dublin, Ireland
e-mail: tom.odwyer@teagasc.ie

T. Skillington and A. Setti (eds.), *Climate Change Resilience Across Societal Contexts*, Energy, Climate and the Environment,
https://doi.org/10.1007/978-3-031-95020-9_16

to maintain wellbeing, adopting to climate change, adopting measures to abate climate and to promote biodiversity. A conclusions section (5) summarises key findings of this chapter.

Irish Agriculture

The agri-food sector accounts for about 7% of economy wide gross value added (GVA) (Central Statistics Office (CSO), 2022). The most recent Census of Agriculture (CSO, 2022), showed that there were 135,037 farms in Ireland with a 278,600 labour force. Farms are mainly family-run farms with an average farm size of 36 ha. The major farming systems in the country are: beef cattle (55%), sheep (13%), dairying (11%), and tillage/crop farms (10%), and other sectors including forestry (11%).

Climate change is already happening and affecting Irish weather and further change is inevitable. Likely outcomes of further warming are: extended growing season, rainfall increasing in likelihood and intensity, prolonged heatwaves and droughts, higher extreme temperatures, fewer cold extremes (Flattery et al., 2024).

The severity of climate change effects, both Internationally and in Ireland, depends on global greenhouse gas emissions reduction targets being met. Ireland's climate change and biodiversity legislation are now considered to provide a context for considering advisory support for farmer wellbeing and climate change mitigation.

Ireland's Climate and Biodiversity Legislation

The Climate Action and Low Carbon Development Act of 2015 (Government of Ireland, 2015), and its subsequent amendment of 2021 (Government of Ireland, 2021), sets out a framework of legally binding targets, introduces carbon budgeting, and sets sectoral emissions ceilings to guide Ireland towards achieving climate neutrality by 2050. The **2015 Act** directs the development of a **National Mitigation Plan** and a **National Adaptation Framework,** while establishing the **Climate Change Advisory Council (CCAC).** The **2021 Amendment Act** strengthens the original legislation by setting more ambitious, legally binding targets. It commits Ireland to achieving a **51% reduction in GHG emissions by 2030,** relative to 2018 levels, and attaining **net-zero emissions by no later than 2050.** The Act also introduces a strengthened governance framework, requiring the government to adopt **five-year carbon** budgets and **sectoral emissions ceilings.**

A central feature of the amended legislation is the implementation of **carbon budgets**, which cap the total amount of GHG emissions permitted over specified five-year periods. The **CCAC** is responsible for proposing these budgets, which are then reviewed and approved by the government. To operationalise the carbon budgets, the government has established **sectoral emissions ceilings,** which set out the maximum GHG emissions permissible for each sector, including agriculture. The agriculture sector has been assigned a ceiling of 106 $MTCO_2e$ for the five-year period 2021 to 2025 (baseline GHG emissions were 23 $MTCO_2e$ in the 2018 baseline year) and a target to reduce its GHG emissions by 25% by 2030, relative to 2018 levels.

The Environmental Protection Agency (EPA) reports GHG emissions on an annual basis, both for the country as a whole, and for each sector of the economy. For 2023 (EPA, 2024a), national GHG emissions declined by 6.8% (v 2022) or 10.7% (v 2018); for agriculture the figures were -4.6% (v 2022) and -2.9% (v 2018). The EPA has also projected that Ireland will reduce GHG emissions by up to 29% by 2030, compared to the target of 51%. Even this would require full implementation of a wide range of policies and plans across all sectors and for these to deliver the anticipated carbon savings (EPA, 2024b).

In summary, Ireland's Climate Action and Low Carbon Development Acts of 2015 and 2021 provide a legal foundation for transitioning to a low-carbon economy. The introduction of carbon budgets and sectoral emissions ceilings represents a strategic approach to emissions management. However, achieving the specified targets will require an increase in the pace and scale of implementation of the various mitigation strategies.

Regarding biodiversity, the 2024 EU Nature Restoration Law sets legally binding targets relating to the restoration of biodiversity including agricultural ecosystems. Ireland's 4th National Biodiversity Action Plan sets the national biodiversity agenda for the period 2023–2030.

However, legislation and policy, while having the object of setting out a vision and setting targets and goals, are, on their own, unlikely to reach the optimum outcome among farmers who are a dispersed and largely self-employed farmer population. Notably, the imposition of regulatory measures may be contested as indicated by EU farmer protests in 2024 related in part to the burden of environmental regulation, which caused the EU Commission to reconfigure its regulatory strategy (Matthews, 2024).

Legislation and policies on climate change can impact farmer's mental health. However, Agricultural Advisory services, known internationally as 'extension' services, support farmers with farm management and adaption related to sustainability and resilience while ameliorating stress, potentially presents a complimentary means of gaining climate change adaption, over time. These two topics are now reviewed, from an Irish perspective, associated with climate change.

Climate Change and Stress

Agriculture is one of the most climate-sensitive sectors and directly affected by changes in physical and chemical climate conditions (Mitter et al., 2019). Farmers are increasingly exposed to climate change and also are part of the solution by adopting mitigation measures by reducing sources or enhancing the capture or storage of greenhouse gases (Balaine et al., 2025). Climate change adaption is multi-faceted involving both physical farm workplace and practice adoption requiring both management and behaviour change. Mitter et al. (2019) state that 'Farmers' adaptation decisions-such as other human behaviour—underlie perceptual and socio-cognitive processes that are interwoven and refer to selecting, organizing, and interpreting sensory information'.

A recent EU-OSHA Report (EU-OSHA, 2022) on mental health in agricultural workers indicates that relatively little is known about the impact of climate change on the mental health of European farmers and farm workers with what research does exist having been conducted in Australia, Canada, and USA. This report draws on limited EU-level research and grey literature showing that unfavourable seasons cause heat exhaustion and increase workloads. The report also indicates that extreme weather events are associated with various mental health outcomes, including depression, hopelessness, trauma, solastalgia, and suicide ideation and occurrence. Thus, legislation and policies around climate change can impact farmer's mental health.

Nye et al (2025) report that the impact of climate change on the mental health of people working in agriculture varies across countries and regions. They report that challenges faced are likely to stem from production issues and anxieties related to identity and sense of place. They also state that the effects of these impacts require immediate examination not only to evaluate levels of preparedness and prescribe adaptation and capacity building interventions required, but also to highlight potential mental health impacts and how these can be negotiated at the varying stages of climate change as it happens.

The Irish study by Balaine et al. (2025) reported that farmers studied were aware of weather changes associated with climate change and consider that a fundamental shift in weather variation has occurred. However, they suggest that Irish weather changes in causing disruption to farming operations impact farmers' mental health and wellbeing due to unpredictability in its occurrence, economic impact, and increased workload (Balaine et al., 2025). It has also been reported in Ireland that farmers often feel subjected to changes outside their control, including legislative and policy requirements of environmental measures and associated administrative procedures, leading to a state of 'liminality' as the source of stress (Firnhaber et al, 2024).

Regarding Irish farmer stress incidence, a large cross-sectional study among both dairy and cattle farmer by van Doorn et al. (2022) found that 12.4% of farmers suffered stress often or very often. Brennan et al. (2021), using a nationally representative sample, found that 57% of farmers experienced 'anxiety or stress' in the previous 5 years, with principal sources being weather related (47%), workload (32%), and financial (28%). More recently, all farmers types, except dairy farmers, reported a significant deterioration in their 'stress' levels over recent years (Dillon, 2024). A previous study among Irish dairy farmers ranked farm stressors in the following order: time pressure; farming bureaucracy; unpredictable events; personal hazards; financial; and isolation (Furey et al., 2016). This study also found that social support significantly reduced mental distress, as a precursor for anxiety, depression, and injury expectation.

A Swedish review concurs with Irish studies, but with inclusion of climate change and weather as a stressor (Lundqvist et al, 2023). Systematic reviews of farmer mental health sources concur broadly with the sources identified in Ireland, with the additional inclusion of pesticide exposure but without inclusion of climate change (Daghagh Yazd et al., 2019; Younker & Radunovich, 2021).

Lane et al. (2019) in NY State USA, however, reported that while dairy farmers articulated concern regarding climate impacts, notably other business pressures, such as profitability, market conditions, government regulations, and labour availability, more often affected their decision-making along with personal experience of extreme weather and seasonal changes.

An Irish review of international literature related to farmers' mental health, health seeking behaviour, and literacy has been undertaken indicating research gaps (O'Shaughnessy et al., 2022). Follow-up studies in Ireland on occupational stressors include sleep issues and burnout associated with excessive workload (O'Connor et al., 2024a); farmer views on current mental health services and preferences for support (Malone et al., 2025); and disordered alcohol and substance use (O'Connor et al., 2024b).

A Swedish study has indicated that farmers may adopt positive or negative strategies for stress management but that resilience can be taught (Lundqvist et al, 2023), which supports advisory service engagement in this aspect of farm management. An EU project has reported that farmer stressors may arise from a range of sources which may or may not be under their control (EU SafeHabitus-Project, 2025).

Agricultural Advisory Services Internationally and in Ireland.

Agricultural Advisory Services, known internationally as 'extension services', assist farmers to develop their farms to remain sustainable in economic, environmental, social, and innovation terms (Paine, 1999). Trust between both an extension agency and individual extension staff members and farmer clients is a vital ingredient to gaining adoption (Bruening et al., 1992). Advisory provision is distinct from regulation described as: control exercised by a public agency over activities that are valued by society as a means of gaining adoption of measure(s) (Gormley, 1998). Considerable regulation of agricultural production, climate change, and occupational safety and health (OSH) has taken place in Europe and Ireland. Advisory Services can play a supporting role in assisting farmers to meet compliance requirements (Gunningham, 2002). A review of the use of advisory service work related to OSH shows that improving farmers' capability to manage OSH is a crucial requirement to gain adoption of OSH measures (McNamara, 2015). However, preventative innovations in public health arenas, such as adoption of health or mental health measures, are essentially voluntary in nature (Rogers, 2003).

A Farm Advisory System (FAS) is operated throughout Europe mandated by a European Parliament Regulation, to assist farmers to meet their obligations under EU supported schemes and programmes. On-going training related to farmer climate change adoption and OSH and health wellbeing promotion is provided to Irish FAS Advisors, which include Teagasc advisors.

A developmental Agricultural Advisory Service is operated in Ireland, named Teagasc—the Irish Agriculture and Food Development Authority (www.teagasc.ie). A feature of this service is a model of integration of Research and Knowledge Transfer (KT) (Knierim & Prager, 2015). Teagasc conducts research on sustainable farming practices, soil carbon sequestration, and emission reduction strategies to provide science-based solutions for farmers. A blueprint for GHG abatement potential of the Irish agriculture and land-use sectors up to 2030, on which advisory services base services, has been published (Lanigan et al., 2023). The

KT component includes advisory services along with both vocational and adult education and training. Principal approaches used include: engagement with farmers individually, at advisory events, meetings, and peer-to-peer discussion groups; use of media including social media (ÓFathartaigh, 2021). Services include farmer support for State scheme and programmes including advice on farm infrastructure modernisation in line with State Department of Agriculture, Food and the Marine (DAFM) specifications, to meet future expected climate change challenges. Modernising infrastructure impacts farmer wellbeing through worktime use and improved OSH and adapts a farm for future climate impacts such as manure storage and storms.

A 'win–win' arises for advisory services by promoting technical efficiency and climate change adaption jointly, as the most economically efficient farms produce the lowest GHG per kilogram or per litre of product (Ryan et al, 2015). Notably, Irish dairy and beef farms have been assessed as having 0.87 and 0.63 technical efficiency coefficients, respectively (Martinez Cillero & Tovar Reaños, 2023) which indicates potential to improve efficiency gain. Research knowledge advances the frontier of technical efficiency giving rise to a constant challenge for both advisors and farmers to adapt to the new knowledge. Technical efficiency also leads to financial sustainability which is an important component of wellbeing and it allows for funding of farm development into the future, including climate change measures.

Current metrics on advancing for advancing innovation, sustainability, technical, and GHG efficiency performance of the Irish agri-food sector are published annually (Teagasc, 2025a).

Teagasc Programme to Promote Farmer Wellbeing.

Teagasc in its advisory programming adopts the Total Worker Health Model (Schill &, Chosewood, 2013) which provides a broad focus on health and wellbeing in conjunction with occupational safety (McNamara et al, 2021). This approach is appropriate as health, including mental health, and safety are interconnected and impact each other while also impacting farm business parameters (McNamara et al, 2007).

Teagasc works in partnership with health promoting agencies in line with health promotion best practice (Corbin et al, 2018). Such partnership approaches allow leverage of the advisory service relationship with farmers to promote farmer health (Hon and Grunig, 1999).

Teagasc promotes farmer health and safety at a national level through publications and media to see attitudinal and behavioural change. Significant publications distributed nationally include: a farmer health booklet (Richardson et al., 2015); farmer mental health booklets (Teagasc/ MHI,); and a health and wellbeing booklet (H.S.A/ F.S.P., 2022).

Advisory services in Ireland can also engage individually with farmers related to health and a State sponsored training programme provides advisors with skills and knowledge to signpost farmers towards health services, particularly mental health. Called 'On Feirm Ground' (meaning 'on solid ground'), the one-day advisor training programme focuses on promoting farmers' health and sets out clear roles, responsibilities, and boundaries for advisors in their signposting role (https://mensnetwork.ie/ofg/). This training programme is based on research into advisors' needs related to health signposting provision (Hammersley et al., 2022). Follow-up evaluation showed a significantly improved advisor participants' self-reported levels of knowledge, capacity, and willingness to support farmer mental health (Hammersley et al. 2024).

Adapting to Climate Change and Advisory Service Provision

While climate mitigation focuses on reducing the sources or enhancing the capture or storage of greenhouse gases, climate adaptation is the process of adjustment to actual or expected climate change (DAFM, 2019). Separate mitigation and adaptation plans for the agriculture sector exist, but, DAFM (2019) highlights that both are part of their overall approach for dealing with climate change to be integrated into the everyday work of DAFM and associated agencies.

The recent study involving Irish farmers (Balaine et al., 2025) highlighted the detrimental impacts of climate events on farmer wellbeing and quality of life. Increased workload, worry, and stress were recorded by farmers dealing with unpredictable weather, associated rising costs, and the uncertainty of the future. While farmers have always adjusted their farm work to weather fluctuations, the study showed that they now face new challenges, with more frequent disruptions to farming operations and resulting environmental, economic, and social impacts. Farmers viewed these changes as a fundamental shift that goes beyond the typical weather variations that they have managed in the past. The same report indicated that farmers understood the need to adopt adaptation practices that reduce the negative impacts of weather changes and increase resilience to future weather events, yet were often uncertain about the type of practices that they could implement. The study also showed that the decision-making associated with farmers' climate adaptation decisions is complex and multi-faceted, influenced by a combination of factors, which at times delayed, and/or added pressure on farmer decision-making.

A New Zealand study (Griffin et al., 2023) found that New Zealand farmers recognise the increasing frequency of extreme weather events but often attribute these to natural variability rather than anthropogenic climate change. It also reported that adaptations are primarily reactive, focusing on immediate challenges like floods or droughts, rather than proactive planning for long-term climatic shifts. Distinctions in adaptive capacity were observed across farmers of different ages, financial status, levels of on-farm intensification, and involvement of family members in farm operations.

Finally, a study involving Welsh farmers (Griffiths & Evans, 2015) found that farmers had a high overall coping capacity and used various strategies to cope with extreme weather events. However, the effectiveness of these strategies varied, and the report concluded by questioning the farmers' ability to enhance their resilience as a specific response to the growing pressures of climate change.

In summary, Irish farmers are increasingly aware of climate change and are adopting various strategies to adapt. However, their perceptions and actions are deeply intertwined with their economic realities, the policy environment, and specific regional contexts, necessitating tailored support and policies to enhance resilience in the agricultural sector. Teagasc has recognised the need to support Irish farmers in both climate mitigation and adaptation, through the development of both new advisory support (including education) services for farmers and the enhancement of its research activities in the climate area. The next section outlines the organisation's role in building farmer capacity to cope with climate change.

Signpost Programme: Supporting Farmers with Climate Change Adaption -

Teagasc collaborates with agri-food industry partners to jointly address challenges and opportunities facing the agricultural industry. An example of this is the Signpost Programme (www.teagasc.ie/signpost). Launched in 2021, it is a multi-annual, whole-of-industry approach to lead and support the transition of Irish farming towards more sustainable farming systems. Its primary objective is to support climate action (reduced GHG emissions) by Irish farmers. However, the programme also promotes climate adaptation, improvements in water quality and biodiversity, while helping farmers to farm profitably. It achieves this through combining the four elements already listed: (1) 'following the science' and recommending science-based climate solutions to Irish farmers; (2) using a range of complementary advisory approaches (including demonstration farms, discussion groups, training workshops, media, social media, on-farm events, one-to-one farmer consultations, and farm visits) to engage with farmers; (3) providing training for young farmers, farmers, and industry/ Agricultural Knowledge Information System (AKIS) actors; and (4) working closely with both policy makers and industry partners.

Promoting Biodiversity

A Teagasc strategic goal is to provide science-based evidence and technologies to enable Irish farmers to meet biodiversity commitments. Following research by Keena (2020), provision of education and advisory services for farmers in the value, maintenance, and enhancement of biodiversity on farms takes place.

Conclusions

With the increasing impact of climate change on farming systems, the need to enhance both farmer resilience and the resilience of farming systems is highlighted. Some farmers, and farming systems, are more resilient to changing weather patterns than others. The ability of farmers to build their resilience is often hampered by factors such as a lack of knowledge, regarding potential solutions, financial constraints, uncertainty about future policy development, and lack of advisory support. These potential barriers to climate adaptation can be lowered by tailored support, enhanced knowledge exchange, financial assistance, and favourable policies. Within Ireland, Teagasc plays a significant role in supporting farmers to adapt to and mitigate climate change through research, advisory services, and education.

More generally, this chapter has described how advisory services in Ireland support farmers with climate change adaptation and mental health resilience by provision of independent and trusted advice. The chapter is framed against an international perspective where advisory services engagement in these issues is urgently required. Drawing on the Socio-ecological Model (Kilanowski, 2017), farmers are centrally located at the centre of the model where their actions are based on their own psychological and managerial characteristics and are influenced by concentric layers including family and friends, farming organisation, advisory services, and legislation and policy measures. Thus, both internal and external stimuli exist to influence farmer actions. Advisors in many respects act in an 'intermediary' role to assist farmers to make short- or long-term science-based decisions, adapted to each farm's individual set of circumstance. Intermediary organisations and persons have been shown to be a powerful ingredient to gain practical change adoption (Sinclair et al, 2013). Advisory services using their expert knowledge and expertise may interact with other actors within the web of influence on farmers (Beseler et al, 2021) to support approaches to make positive progress with issues such as climate change adaption and wellbeing.

References

Balaine, l., Henchion, M., Á Macken-Walsh, Á., Meredith, D., Mohammadrezaei, M., & Sharon Sweeney, S. (2025). Farm-level behavioural change towards building climate resilience: Insights from a mixed-methods study. Irish Climate Change Advisory Council (Working Paper Series Number 36). https://www.climatecouncil.ie/councilpublications/councilworkingpaperseries/Farm-level%20behavioural%20change%20towards%20building%20climate%20resilience%20publication.pdf

Beseler CL, Keeney AJ, Garratt R, Wertheimer A. & Stallones L. (2021, April). Assessing connections in an agricultural community using social network analysis. *Journal of Agromedicine, 26*(2), 245–255. https://www.tandfonline.com/doi/full/https://doi.org/10.1080/1059924X.2020.1837317

Brennan, M., Hennessy, T., Meredith, D., & Dillon, E. (2021). Weather, workload and money: Determining and evaluating sources of stress for farmers in Ireland. *Journal of Agromedicine, 27*(2), 132–142. https://doi.org/10.1080/1059924X.2021.1988020

Bruening, T. H., Radhakrishna, R. B., & Rollins, T. J. (1992). Environmental issues: Farmers' perceptions about usefulness of informational and organizational sources. *Journal of Agricultural Education, 33*(2), 34–42. https://doi.org/10.5032/jae.1992.02034

Central Statistics Office (2022). *Census of agriculture 2020, detailed results – Press Release*. https://www.cso.ie/en/releasesandpublications/ep/p-coa/censusofagriculture2020detailedresults/

Corbin, J. H., Jones, J., & Barry, M. M. (2018). What makes intersectoral partnerships for health promotion work? A review of the international literature. *Health Promotion International, 33*(1), 4–26. https://academic.oup.com/heapro/article/33/1/4/2555416?login=true

Daghagh Yazd, S., Wheeler, S. A. & Zuo, A. (2019). Key risk factors affecting farmers' Mental health: A systematic review. *Int J Environ Res Public Health,16*(23), 4849. https://pubmed.ncbi.nlm.nih.gov/31810320/

Dillon, E. (2024) *Teagasc National Farm Survey 2023 Sustainability Report (2024)*. Social Sustainability Indicators (pp. 68–70). https://www.teagasc.ie/media/website/publications/2024/2023-Sustainability-Report.pdf

Department of Agriculture, Food and the Marine (DAFM, 2019) https://assets.gov.ie/220412/8eb5a3c1-cde7-469d-bd5b-2d022b394dc2.pdf What ?

Environmental Protection Agency (EPA) (2024a). https://www.epa.ie/news-releases/news-releases-2024/irelands-greenhouse-gas-emissions-in-2023-lowest-in-three-decades-.php. Accessed March 1, 2025.

Environmental Protection Agency (EPA, 2024b) https://www.epa.ie/news-releases/news-releases-2024/ireland-is-projected-to-exceed-its-national-and-eu-climate-targets.php. Accessed March 1, 2025.

European Union Horizons Project: SafeHabitus (2025, March 13). Deliverable 4.1. Current and future drivers of agricultural change: Assessing the implications for working conditions and occupational stress amongst farmers and farm workers. https://www.safehabitus.eu/wp-content/uploads/2024/08/SafeHabitus_D4.1.pdf

European Agency for Safety and Health at Work (2022). *Mental health in agriculture: preventing and managing psychosocial risks for farmers and farm workers Report*. https://osha.europa.eu/sites/default/files/documents/Mental-health-agriculture_EN.pdf

Firnhaber, J., Malone, S. M., O'Hagan, A. D., O'Keeffe, S., McNamara, J., & O'Connor, S. (2024). "People that are supporting [the] whole sector are on their knees"; Uncertainty and socioeconomic change are occupational stressors for Irish farmers. *Rural Sociology (USA), 89*(3),483–505. https://onlinelibrary.wiley.com/doi/full/https://doi.org/10.1111/ruso.12554

Flattery, P. (2024). *How weather patterns are changing, what are we adapting to?*. Contained in Conference Proceedings of Climate Adaptation Conference. https://www.teagasc.ie/media/website/publications/2024/Climate-adaptation-conference-proceedings.pdf

Furey, E. M., O'Hora, D., McNamara, J., Kinsella, S., & Noone, C. (2016). The roles of financial threat, social support, work stress and mental distress in dairy farmers' expectations of injury. *Published in Frontiers in Public Health, Occupational Health and Safety, an Online Open Access Journal and is Available at*. https://doi.org/10.3389/fpubh.2016.00126

Government of Ireland. (2015). *Climate action and low carbon development Act 2015*. https://www.irishstatutebook.ie/eli/2015/act/46/enacted/en/html

Government of Ireland. (2021). *Climate action and low carbon development (Amendment) Act 2021*. https://www.irishstatutebook.ie/eli/2021/act/32/section/15/enacted/en/html

Griffin, C., Wreford, A., & Cradock-Henry, N. A. (2023). 'As a farmer you've just got to learn to cope': Understanding dairy farmers' perceptions of climate change and adaptation decisions in the lower South Island of Aotearoa-New Zealand. *Journal of Rural Studies, 98*, 147–158.

Griffiths, R., & Evans, N. (2015). The Welsh Marches: resilient farmers? Exploring farmers' resilience to extreme weather events in the recent past. *Journal of Depopulation and Rural Development Studies, 18*, 161–189.

Gormley T.F. Jr. (1998). Regulatory enforcement styles. *Political Research Quarterly, 51*(2), 363–383. https://www.proquest.com/docview/215324235?sourcetype=Scholarly%20Journals

Gunningham, N. (2002, March). Regulating farm safety: Towards an optimal policy mix (pp. 2–15). National Research Centre for OHS Regulation. https://openresearch-repository.anu.edu.au/server/api/core/bitstreams/6d6edebd-c461-4783-ac1e-ed6f8acff038/content

Hammersley, C. R., Meredith, M., Carroll, P., & McNamara, J. (2022). Supporting farmer wellbeing: Exploring a potential role for advisors. *The Journal of Agricultural Education and Extension, 29*(4), 511–538. https://doi.org/10.1080/1389224X.2022.2082498

Hammersley, C., Richardson, N., Meredith, D., McNamara, J. Carroll, P., & Jenkins P. (2024). 'On Feirm Ground': analysing the effectiveness of a bespoke farmer mental health training programme targeted at farm advisors in Ireland. *The Journal of Agricultural Education and Extension*, 1–29 https://www.tandfonline.com/doi/full/https://doi.org/10.1080/1389224X.2024.2339801

Health and Safety Authority/ Farm Safety Partnership Booklet (2022). *Farmers' Health and wellbeing. A guide to staying healthy while farming*. https://www.hsa.ie/eng/publications_and_forms/publications/agriculture_and_forestry/farmers_health_and_wellbeing.html

Hon, L. C., & Grunig J. E. (1999). *Guidelines for measuring relationships in public relations*. The Institute for Public Relations, Gainesville. Online.at: http://www.instituteforpr.com.

Keena, C. (2020). *An examination of biodiversity management practices on Irish farms and how this can be measured: the case of dairy farmers in County Waterford*. PhD Thesis National University of Ireland, University College Dublin.

Kilanowski, J. F. (2017). Breath of the socio-ecological model. *Journal of Agromedicine, 22*(4), 295–297. https://doi.org/10.1080/1059924X.2017.1358971

Knierim, A., & Prager, K. (2015). *Agricultural knowledge and information systems in Europe: Weak or strong, fragmented or integrated?*https://430a.unihohenheim.de/fileadmin/einrichtungen/430a/PRO_AKIS/About/OVERVIEW.OF.AKIS.IN.EUROPE.AKIS_characterisation_briefing_final.pdf

Lane, D., Murdock, E., Genskow, K., Rumery Betz, C., & Chatrchyan, A. (2019). Climate change and dairy in New York and Wisconsin: Risk perceptions, vulnerability, and adaptation among farmers and advisors. *Sustainability, 11*, 3599. https://doi.org/10.3390/su11133599

Lanigan, G., Black, K., Donnellan, T.,Crosson, P.,Beausang, C., Hanrahan, K., Buckley, C., Lahart, B., Herron, J., Redmond, J., Shalloo, L., Krol, D., Forrestal, P., Farrelly, N., O'Brien, D., Lenehan, JJ., Hennessy, M., O'Donovan, M., Wall, D., O'Sullivan, L., O'Dwyer, T., Dineen, M., Waters, S., NiFlahartha, N., Houlihan,T., Murphy, P., Spink, J., Dillon, P., Upton, J., & Richards, K. (2023). *MACC 2023: An updated analysis of the greenhouse gas abatement potential of the Irish agriculture and land-use sectors between 2021 and 2030* (pp. 353pp). Teagasc. https://www.teagasc.ie/media/website/environment/climate-action/climate-centre/MACC-2023.pdf

Lundqvist, P., Håkansson, C., & Hakelius, K. (2023). *Farmers' work environment and mental health. Publisher: Swedish agency for work environment expertise*. Number: 2022:8. https://publications.slu.se/?file=publ/show&id=121646

Malone, S. M., Firnhaber, J., O'Hagan, A. D., O'Keeffe, S., McNamara, J., & O'Connor, S. (2025). "Lacking the rural empathy"; Irish farmers' and stakeholders' opinions on current mental health services and preferences for support. *Journal of Rural Studies,* 114. https://doi.org/10.1016/j.jrurstud.2024.103508

Martinez Cillero, M., & Tovar Reaños, M. (2023). Farm technical and environmental efficiency and subsidy redistribution in Ireland: A simulation approach of possible performance and equity effects. *Journal of Agricultural Economics, 74*, 394–412. https://doi.org/10.1111/1477-9552.1250

Matthews, A. (2024). Farmer protests and income developments in the EU. *The Political Quarterly, 95*, 344–349. https://doi.org/10.1111/1467-923X.13396

McNamara, J. (2015). *A study of the effectiveness of risk assessment and extension supports for Irish farmers to improve farm safety and health management*. Unpublished PhD Thesis. University College Dublin. https://t-stor.teagasc.ie/handle/11019/1840

McNamara, J., Fox, M., Kinsella, J., & O'Connor, D. (2021). Promoting Farmer occupational safety and health (OSH) services through Extension. *Agronomy Research, 19*(S2), 1060–1074, https://doi.org/10.15159/AR.21.059

McNamara, J., Ruane, D. J., Whelan, S., & Connolly, L. (2007). A preliminary investigation of the incidence and impact of disability on Irish farms. *Journal of International Agricultural and Extension Education, 14*(2), 21–32.

Mitter, H., Larcher, M., Schönhart, M., Stöttinger, M., & Schmid, E. (2019) . Exploring farmers' climate change perceptions and adaptation intentions: Empirical evidence from Austria. *Environmental Management, 63*, 804–821. https://doi.org/10.1007/s00267-019-01158-7

Nye, C., Wheeler, R., Rose, D., Becot, F., Holton, M., Hocevar, D. K., Knook, J., Kyle, S., Partalidou, M., Riley, M., Steiner, A., & Whitley, H. (2025). Mental health, well-being and resilience in agricultural areas: a research agenda for the Global North. *Journal of Rural Studies, 114*, Article 103506. https://doi.org/10.1016/j.jrurstud.2024.103506

Paine, M. S. (1999). Improving the management of technology development through mediation. *The Journal of Agricultural Education and Extension, 6*(2), 93–110.

O'Connor, S., Malone, S. M., Firnhaber, J., O'Shaughnessy, B. R., McNamara, J., & O'Hagan, D. (2024). Disordered alcohol and substance use in Irish farmers: A cross-sectional survey. *The Journal of Rural Health, 40*(1), 173–180.

O'Connor, S., O'Hagan, A. D., Malone, S. M., O'Shaughnessy, B. M., McNamara, J., & Firnhaber , J. (2024). Sleep issues and burnout in Irish farmers: a cross sectional survey. *Safety Science*, 171. https://www.sciencedirect.com/science/article/pii/S0925753523003193

ÓFathartaigh, M. (2021). *Developing rural Ireland. A history of the Irish agricultural advisory services*. Publisher Wellword. ISBN: 9781913934606

O'Shaughnessy, B. R., O'Hagan, D., O'Connor, S. (2022). *Review of literature on farmers' mental health*. A Report funded by DAFM. Dublin City University.

Richardson, N., Osborne, A., O'Neill, B., Griffin,P., McNamara, J. Roche,C., & van Doorn, D. (2015) Staying fit for farming—A health booklet designed for Irish farmers. *Journal of Agromedicine, 20*(3), 381–385. https://www.tandfonline.com/doi/full/https://doi.org/10.1080/1059924X.2015.1047551

Ryan, M., Buckley, C., Dillon, E. J., Donnellan, T., Hanrahan, K., Hennessy, T., Moran, B., & Kelly, E. (2015). The most economically efficient farms produce the lowest GHG emissions per Kg of Product. *EuroChoices, 14*(1), 26–27.

Rogers E.M., (2003). Diffusion of innovations (5th Edn, p. 367). Free Press.

Schill, A., Chosewood, L. C. (2013). The NIOSH total worker health™ program: An overview. *Journal of Occupational and Environmental Medicine, 55*, S8-S11. https://journals.lww.com/joem/fulltext/2013/12001/The_NIOSH_Total_Worker_Health__Program__An.2.aspx

Sinclair, R. C., Cunningham, T. R., & Schulte, P. A. (2013). A model for occupational safety and health intervention diffusion to small businesses. *American Journal Industrial Medicine, 56*(12), 1442–51. https://onlinelibrary.wiley.com/doi/https://doi.org/10.1002/ajim.22263

Teagasc/Mental Health Ireland Booklet (2019). *Coping with 'the pressures of farming'*. https://www.mentalhealthireland.ie/wp-content/uploads/2020/04/Coping-with-the-pressures-of-farming.pdf

Teagasc/Mental Health Ireland Leaflet (2024). *Sowing the seeds of support*. Positive Mental Health Guidance for the Farming Community. https://www.teagasc.ie/publications/2024/sowing-seeds-of-support-positive-mental-health-guidance-for-the-farming-community.php

Teagasc (2025a). *Advancing innovation, sustainability and technical performance of the agri-food sector in 2024*. https://www.teagasc.ie/publications/2025/advancing-innovation-sustainability-and-technical-performance-of-the-agri-food-sector-in-2024.

Teagasc, (2025b). *Teagasc farm health and safety programme*. In Teagasc Programme submitted to DAFM.

Van Doorn, D., Richardson, N., Meredith, D., McNamara, J., & Blake, C. (2022). *Farmers have hearts cardiovascular health programme detailed impact report*. https://www.teagasc.ie/media/website/publications/2022/Farmers-Have-Hearts---Cardiovascular-Health-Programme---Detailed-Impact-Report.pdf

Younker, T., & Radunovich, H. L. (2021). Farmer mental health interventions: A systematic review. *Int J Environ Res Public Health, 26, 19*(1), 244. https://www.mdpi.com/1660-4601/19/1/244

17
Conclusion

Tracey Skillington and Annalisa Setti

Learning from diverse communities' strategies for minimizing the impacts of climate change and strengthening the co-benefits of planetary and human wellbeing has helped to visualize sustainable futures in more inclusive and proactive terms. However, we still face considerable challenges when attempting to put new resilience-building measures into practice. What have become increasingly clear in recent years are the ways climate change is altering basic ecologies of place to a point where place attachments and identities are forcibly redefined, especially in scenarios of large-scale change (e.g., the Arctic and Antarctica, see Rantanen et al., 2022). As a result, displacement occurs on multiple levels—geographically in response to disaster events, socially with the

T. Skillington (✉)
University College Cork, Cork, Ireland
e-mail: t.skillington@ucc.ie

A. Setti
Psychology Enterprise Centre, University College Cork, Cork, Ireland
e-mail: a.setti@ucc.ie

T. Skillington and A. Setti (eds.), *Climate Change Resilience Across Societal Contexts*, Energy, Climate and the Environment,
https://doi.org/10.1007/978-3-031-95020-9_17

loss of one's social role or 'place in the world', and psycho-culturally with the disappearance of knowledge or skill sets linked to a specific environment or place. To build truly effective resilience programme, therefore, requires that all of these forms of displacement are addressed. However, one must first comprehend how displaced communities interpret what is lost through these change processes as lived experiences.[1] Place-based loss, for instance, is a common source of climate change-related loss examined throughout the chapters of this book (e.g., loss of home, viable farmlands, accessible, healthy city spaces, etc.).

Bonds of attachment to place and the type of historical, social, cultural, psychological, and economic values invested in them have been shown to centrally shape willingness to implement protective measures and strengthen preparedness for disaster (e.g., flooding or wildfires) (e.g., see Adger et al., 2013; Anton & Lawrence, 2016; Cunsolo et al., 2012; Mishra et al., 2010; Scannell & Gifford, 2013). When bonds of attachment are disrupted by the effects of climate change (e.g., the slow erosion of viable farmlands and traditional practices, and relatedly, community identities, emotional security, and imagined futures), it can seriously undermine the resilience of communities. The intertwining of landscapes, ecosystems, and climate with culture, history, social, psychological, and physical health connections powerfully shapes everyday ways of seeing, engaging, and understanding effective strategies for adapting to a changing world (Creswell, 2004; Hess, 2008). Recent longitudinal studies of policy approaches to climate mitigation and resilience-building have found the vast majority to be of limited success. One study by Stechemesser et al. (2024), for instance, evaluated 1500 climate policies in forty-one countries between the years 1998 and 2022 and found just 63 policies contributed successfully to a stable reduction in pollution levels and a notable improvement in local environmental conditions.[2] The vast majority of climate policies, as this and similar studies show, are failing in their primary aim—to protect the self-determining capacities of communities exposed to conditions that increasingly undermine both personal and collective lives.

Agency not only forms the basis of pre-emptive engagements with a climate changing world but is also crucial to maintaining one's status as a self-determining subject. Exposure to a growing range of sources of climate inspired loss can undermine peoples' sense of agency with respect to the shaping of one's destiny, autonomy, and self-governing capacities. Efforts to address the psychological dimensions of these losses have gained momentum in recent decades (e.g., on 'climate anxiety', see the NRDC, 2024; on 'solastalgia', see Albrecht et al., 2007; McManus et al., 2014) and while attention to climate smart infrastructure, heatwave, and flooding resilient city planning has also improved considerably (e.g., see UNEP, 2021; UK Green Building Council, 2025), less well researched is the question of how climate losses and damages might be addressed politically in ways that give voice to differing experiences of loss and evaluations of resilience-building efforts. The political dimensions of climate resilience-building remain poorly understood.

Resilience-building, arguably, ought to be interpreted as an ongoing political, social, and individual endeavor, as much as a prescriptive policy procedure. Continuous negotiations of the meaning and consequences of our changing climate condition are an essential interpretive accomplishment of communicative action and democratic decision-making (Habermas, 1984: 216; O'Mahony et al., 1996) yet are clearly lacking key elements at present. Most notably, deliberative principles of mutual understanding and reciprocity (in terms of a more active support for the exploration of differences in values, perspectives, preferences, and beliefs) should be promoted. Apart from the scientific facts of climate-related losses, consideration must also be given to the issue of values, identity concerns, and creating shared, mutually intelligible fair standards of environmental care that respect history and avoid culturally biased perspectives on sustainable development. There is also a need to reflect openly on the political nature of these issues. 'Green transitions' are rarely straightforward and non-controversial but heavily influenced by issues of inequality, power, and political authority. The discounting of contributions from youth, older persons, the neurodivergent, differently abled peoples, Indigenous, or marginalized socio-economic communities has been widespread and takes several forms, including testimonial injustice, where the experiences and views of these different groups are downplayed, or minimum effort is made to make sense of their understandings of loss (hermeneutic injustice) and incorporate them into mainstream policies and practice. Instead, the translation of the values, knowledge skills, and beliefs of these groups into resilience-building societal programmes is routinely suppressed (e.g., Indigenous non-siloed approaches to nature and culture, see Vizenor, 2008). Unequal access to essential components of resilience-building plans has weakened overall efforts at climate mitigation and adaptation to date, as well as the democratic potential of decision-making arrangements.

More critical attention to such failures as the product of political choices is needed. In particular, the delimitation of the content of deliberations on resilience-building to serve selective ends (i.e., why are some resilience pathways prioritized over other possibilities, who decides what measures will be taken and why). Trade-offs between values and priority concerns rarely occur in a political void. Which values should be regarded as the most pertinent and acted upon depends on the communicative and political institutional power of the actors involved. It is essential, therefore, to look at the political point of democratic deliberation on climate futures and conceive of the question of justification (e.g., for high-risk technological options) more openly as a practical concern requiring immediate attention.

If climate mitigation and resilience-building are genuine end goals, the reasons why current approaches can be considered valid by all need to be the subject of a wider critical debate where the possibility of rejecting or revising established procedures can be a viable option (Forst, 2014). To further this process, we propose the establishment of a series of officially endorsed public hearings recounting experiences of climate-related loss and resilience-building efforts across a range of issues among a diverse gathering of community representatives.[3] These hearings would then be followed by a series of deliberative forums outlining new cooperative efforts to help fellow communities strengthen their resilience capacities (e.g., by ensuring more equitable long-term resource management arrangements, risk assessment, and decision-making procedures are established) followed by a joint event with government officials, community, and civil society organizations to ensure that new initiatives and concerns feed into policy revisions and subsequent climate actions.

Arguably, such procedures are necessary to protect the 'knowledge sovereignty' of all communities and ensure diverse models of sustainable living are respectfully preserved and envisaged for future generations.

Equally, to advance understanding of how climate losses and damages are generated collectively and how effective responses to the same necessitate joint action. Further deliberation is also needed to explore particularity and difference in evaluations of loss and effective response systems (i.e., context specific) and to consider what might constitute an equitable trade-off or integration of the varied needs and values of different groups or communities might best be determined? What is essential is that the particular history and social position of each participating group in this deliberation process are recognized in dialogue with others and in ways that contribute to an 'enlarged' (Benhabib, 1991) perspective on the question of resilience. The inherent diversity and fluidity of social worlds demand that we move away from a static, 'one size fits all' model of resilience in climate negotiations, even if a convergence on aims (e.g., peace and security) remains a regulative ideal of global cooperation.

A deliberatively grounded, relational framing of climate change losses and their prevention starts from the understanding that 'our common future' (World Commission on Environment & Development, 1987) is chiefly determined in the spaces between us, where the activities of multiple agents shape the lives of all, as well as the capacities of each to flourish in a climate changing world (Skillington, 2023: 42). The relational nature of shared planetary wellbeing makes questions of agency, responsibility, cooperation, and justice key to the survival of interdependent, resilient individuals and wider communities. For instance, deliberation on shared responsibilities for the shaping of common futures must connect meaningfully with a discourse on the rights of overlapping communities to ensure that basic freedoms, capabilities, and opportunities to become more resilient are protected over time. Potentially, in their capacity to improve the fluency of democratic conversations on resilience-building, deliberative forums can help bring such issues into a more direct communication with the lived experiences of subjects and encourage a thinking together about shared concerns (Dryzek & Niemeyer, 2024), as much as the peculiar challenges each community face. How conversations on resilience-building can be scaled up to inform large-scale policy also requires critical attention in light of the evolution of media usage and grass root movement participation.

However, presently we lack the type of deliberative mechanisms needed to address such issues as issues of non-recognition and radical inequality experienced by those living in a climate age 'dictated by others' (Auyero & Swistun, 2009: 129). That is, those not rendered fully present in dominant visions of resilient futures or structurally positioned to empower alternatives to 'walls, guns, and finance' (Wainright & Mann, 2018: 197) when addressing mounting ecological concerns. Instead, the demands of the current global economic order are thought to compel a wholesale commitment to a model of expansion at all costs, a stance marked by its unimaginativeness, as much as its apolitical vigor (Skillington, 2023). The current impasse in climate change mitigation is, among other things, 'a crisis of the imagination' yet it is precisely imagination that is needed to inspire a more creative institutional relationship with climate futures and the human potential for change (see Veale, O'Rourke & Byrne this volume; Tavory & Eliasoph, 2013; Skillington, 2024). Effective resilience-building requires publics to embrace at least three things. Firstly, the courage to square up to the worst elements of our predicament by working through the despair and destruction occasioned by the growing impact of climate change on everyday lives, especially as they affect the most vulnerable. Second, by embracing the human capacity for autonomy, commit to actions that explore and democratically nurture alternatives at our disposal (Jaeggi, 2014: 204; Skillington, 2015). Thirdly, to foster through these actions a way of living that aims to establish co-benefits instead of trade-offs (Setti & MacIntyre, 2023) and encourage a more radical form of hope (Lear, 2006: 121). That is, a hope that comes from a clear understanding of climate risks and the actions, in psychological and practical terms, that need to be undertaken for diverse societies to remain resilient and inclusive.

Notes

1. As the chapters of this volume discuss at length, climate-related losses include loss of knowledge of a sense of place, belonging, social cohesion, identity, and futures. That is, losses that on account of their more nuanced, socio-culturally charged nature are

often omitted from official loss databases. Equally, losses incurred through slow-onset incremental changes tend to be disregarded.

2. The UNFCCC 'L&D' (i.e., losses and damages) approach, for instance, see climate change losses as largely irreversible in the sense that restoration is not possible. Damages, by contrast, are thought to be potentially reversible through resilience-building and restoration projects. The UNFCCC's Warsaw International Mechanism, however, sees such efforts as highly dependent on the effectiveness of international negotiations and an increase in support for climate mitigation and adaptation policies which have not been straightforward to date. Also, the creation of the WIM programme signals that the UNFCCC foresees significant failure in both resilience-building and adaptation across international contexts in the coming years, particularly in poorer, developing regions (Verheyen, 2012).
3. We take inspiration here from the recent public hearings (January and February 2025) organized by the World Health Organization with peoples, organizations, and experts from across Europe who came together to discuss to share insights and concerns on what needs to be prioritized in Europe's new health agenda for 2030 and beyond (WHO, 2025).

References

Adger, W. N., Barnett, J., Brown, K., et al. (2013). Cultural dimensions of climate change impacts and adaptations. *Nature Climate Change, 3*(2), 112–117.

Albrecht, G., Sartore, G. M., Connor, L., Higginbotham, N., Freeman, S., Kelly, B., Stain, H., Tonna, A., & Pollard, G. (2007). Solastalgia: The distress caused by environmental change. *Australasian Psychiatry, 15*(Suppl 1), S95–S98.

Anton, C. E., & Lawrence, C. (2016). Does place attachment predict wildfire mitigation and preparedness? A Comparison of Wildland- Urban Interface and Rural Communities. *Environmental Management, 57*(1), 148–162.

Auyero, J., & Swistun, D. A. (2009). *Flammable: Environmental suffering in Argentine Shantytown.* Oxford University Press.

Benhabib, S. (1991). *Situating the self*. Routledge.

Creswell, T. (2004). Defining place. In T. Cresswell (Ed.), *Place: A short introduction.* Blackwell.

Cunsolo Willox, A., Harper, S. L., Ford, J. D., Landman, K., Houle, K., & Edge, VI. (2012). "From this place and of this place": Climate change, sense of place and health in Nunatsiavut, Canada'. *Social Science & Medicine, 75*(3), 538–47.

Dryzek, J., & Niemeyer, S. (2024). How to constitute global citizens' forums: Key selection principles'. *Global Politics, 15*(4), 604–614.

Forst, R. (2014). *Justification and critique.* Polity Press.

Habermas, J. (1984) *The theory of communicative Action, Vol. 1, reason and the rationalisation of society.* Cambridge: Polity Press.

Hess, J. J. (2008). Climate change: The importance of place. *American Journal of Preventative Medicie, 35*(5), 468–478.

Jaeggi, R. (2014). *Alienation.* Columbia University Press.

Lear, J. (2006). *Radical Hope. Ethics in the face of cultural devastation.* Harvard University Press.

McManus, P., Albrecht, G., & Graham, R. (2014). Psychoterratic geographies of the Upper Hunter region, Australia. *Geoforum, 51*, 58–65.

Mishra, S., Mazumdar, S., & Suar, D. (2010). Place attachment and flood preparedness. *Journal of Environmental Psychology, 30*(2), 187–197.

Natural Resources Defence Council (NRDC) (2024, June 25). *Its time to talk about climate anxiety.*Available at: https://www.nrdc.org/stories/its-time-talk-about-climate-anxiety

O'Mahony, P., et al. (1996). Sustainable development as an organising principle for discursive democracy? *Sustainable Development, 4*(1), 42–51.

Rantanen, M., Karpechko, A. Y., Lipponen, A., et al. (2022). The Arctic has warmed nearly four times faster than the globe since 1979. *Communications, Earth & Environment, 3*(168), 1–10. https://doi.org/10.1038/s43247-022-00498-

Scannell, L. G., & R. (2013). Personally relevant climate change: The role of place attachment and local versus global message framing in engagement. *Environment and Behaviour, 45*(1), 60–85.

Setti, A, and T. MacIntyre. (2023). Person-nature fit: Fostering well-being through nature. In J Burke, Boniwell S, Frates E, O'Boyle C and Lianov L. (Eds) *Routledge international handbook of positive health sciences: Positive*

psychology and lifestyle medicine research, theory and practice(pp. 237–251). Routledge.
Skillington, T. (2023) A relational view of responsibility for climate change effects on the territories and communities of the Arctic. In Arctic justice: Environment, society & governance (pp. 36–50). Bristol University Press.
Skillington, T. (2024). A negative dialectics of climate change destruction: Reflecting on climate futures with Adorno. *Irish Journal of Sociology, 32*(3), 242–251.
Skillington, T. (2015). Theorizing the anthropocene. *European Journal of Social Theory, 18*(3), 229–235.
Stechemesser, A., et al. (2024). Climate policies that achieved major emission reductions: Global evidence from two decades. *Science, 385*(6711), 884–892.
Tavory, I., & Eliasoph, N. (2013). Coordinating futures: Toward a theory of anticipation. *American Journal of Sociology, 118*(4), 908–942.
UK Green Building Council (2025) How can we adapt our buildings to climate change? Available at: https://ukgbc.org/our-work/topics/resilient-buildings/how-can-we-adapt-our-buildings-to-climate-change/.
United Nations Environment Programme. (2021). *A practical guide to climate resilient buildings and communities*. Available at: https://www.unep.org/resources/practical-guide-climate-resilient-buildings.
Verheyen, R. (2012) *Tackling loss & damage—A new role for the climate regime*. Loss and Damage in Vulnerable Countries Initiative.
Vizenor, G. (2008). Aesthetics of survivance. In G. Vizenor (ed.) Survivance: Narratives of native presence (pp. 1–25). University of Nebraska Press.
Wainwright, J., & Mann, G. (2018). *Climate Leviathan: A political theory of our planetary future*. Verso.
World Commission on Environment and Development. (1987). Report of the world commission on environment and development: Our common future. Available at: https://www.are.admin.ch/are/en/home/media/publications/sustainable-development/brundtland-report.html.
Worl Health Organisation (2025) *Climate change and health: Shaping a resilient future*. Available at: https://www.who.int/europe/news-room/events/item/2025/02/28/default-calendar/climate-change-and-health-shaping-a-resilient-future#:~:text=28%20February%202025%2C%2010%3A00,heatwaves%20and%20wildfires%20each%20year.

9783031950193